048342

ECONOMIC AND ECOLOGICAL IN

Brunel University Library
TWICKENHAM CAMPUS
Renewals 0181 744 2869

DATE DUE

29 JAN 1999

EX LIBRIS

Maria Grey Libr...
300 St. Margaret's Road, ...kenham

304.2
301.31
ORG

048342

ORGANISAT ... NT

P20/35

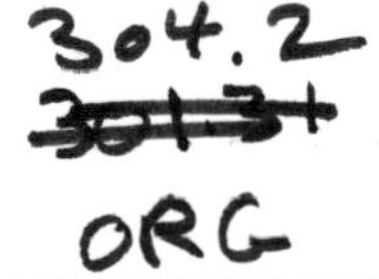

The Organisation for Economic Co-operation and Development (OECD) was set up under a Convention signed in Paris on 14th December 1960, which provides that the OECD shall promote policies designed:

— to achieve the highest sustainable economic growth and employment and a rising standard of living in Member countries, while maintaining financial stability, and thus to contribute to the development of the world economy;
— to contribute to sound economic expansion in Member as well as non-member countries in the process of economic development;
— to contribute to the expansion of world trade on a multilateral, non-discriminatory basis in accordance with international obligations.

The Members of OECD are Australia, Austria, Belgium, Canada, Denmark, Finland, France, the Federal Republic of Germany, Greece, Iceland, Ireland, Italy, Japan, Luxembourg, the Netherlands, New Zealand, Norway, Portugal, Spain, Sweden, Switzerland, Turkey, the United Kingdom and the United States.

304.2
ORG
HC
79
E5
E282
048342

Publié en français sous le titre :

INTERDÉPENDANCE
ÉCONOMIQUE ET ÉCOLOGIQUE

© OECD, 1982
Application for permission to reproduce or translate all or part of this publication should be made to:
Director of Information, OECD
2, rue André-Pascal, 75775 PARIS CEDEX 16, France.

This report is published under
the responsibility of the Secretary-General of the OECD.
It reflects the views of experts speaking on their own behalf and not on behalf
of their Governments, and does not necessarily
represent the views of OECD.

Also available

THE ENVIRONMENT CHALLENGES FOR THE 1980s (October 1981) "Document" Series
(97 81 07 1) ISBN 92-64-12249-4 52 pages £2.70 US$6.00 F27.00

ENVIRONMENT POLICIES FOR THE 1980s (May 1980)
(97 80 02 1) ISBN 92-64-12049-1 110 pages £4.00 US$9.00 F36.00

THE COSTS AND BENEFITS OF SULPHUR OXIDE CONTROL: A Methodological Study (March 1981)
(97 81 01 1) ISBN 92-64-12151-X 166 pages £4.80 US$12.00 F48.00

TRANSFRONTIER POLLUTION AND THE ROLE OF STATES (June 1981) "Document" Series
(97 81 03 1) ISBN 92-64-12197-8 204 pages £5.00 US$12.50 F50.00

COMPENSATION FOR POLLUTION DAMAGE (August 1981) "Document" Series
(97 81 04 1) ISBN 92-64-12214-1 210 pages £5.40 US$13.50 F54.00

EMISSION CONTROL COSTS IN THE TEXTILE INDUSTRY (March 1981) "Document" Series
(97 80 08 1) ISBN 92-64-12134-X 180 pages £4.20 US$10.50 F42.00

MARIA GREY
LIBRARY
W.L.I.H.E.

Prices charged at the OECD Publications Office.

THE OECD CATALOGUE OF PUBLICATIONS and supplements will be sent free of charge on request addressed either to OECD Publications Office, 2, rue André-Pascal, 75775 PARIS CEDEX 16, or to the OECD Sales Agent in your country.

TABLE OF CONTENTS

FOREWORD

In April 1981, the OECD's Environment Committee held a Special Session on "The OECD and Policies for the '80s to Address Long-Term Environment Issues". Marking the beginning of the Committee's second decade of work, the Session brought together four distinguished speakers who launched a discussion of global resource and environment issues within the Environment Committee and throughout the OECD.(1)

Thereafter, at the request and with the support of Member governments, the Secretariat sought the advice of experts on a range of questions concerning these issues, including:

- What are the most critical of the long-term issues on resources and the environment?
- To what extent have scientists succeeded in securing consensus on the causes and effects, nature and consequences of these issues? Do large areas of uncertainty remain to be clarified?
- To what extent do these issues derive from activities of OECD member countries?
- What kinds of policies are now or may in the future be required to address them?

---

(1) See: "The Environment: Challenges for the '80s". Proceedings of a Special Session of the OECD Environment Committee held on April 1st 1981, including statements by: Mr. E. van Lennep, the Secretary-General of the OECD; Mr. Saburo Okita, former Japanese Foreign Affairs Minister and author of the report on "Basic Directions in Coping with Global Environmental Problems"; Mr. Harlan Cleveland, Advisor on the Global 2000 Report to the U.S. President and currently Director of the Hubert Humphrey Institute of Public Affairs; and Mr. Jacques Lesourne, Professor of Economics at France's Conservatoire National des Arts et Métiers and former Director of OECD's Interfutures Project.

- What are the major constraints to, and opportunities for, research or policy action on these issues?

- What work is now under way or planned at the international level on these issues, recognizing that most work is and will continue to be done at the national level?

- What additional work needs to be supported through international co-operation?

The papers published in this volume emerged largely from a consideration of these questions. Each was subjected to several rounds of reviewing and editing by experts. Two special meetings were convened by the Secretariat to consider them. An expert's meeting in July advised on the identification of the issues and undertook an initial review of each. In October, a larger workshop was convened by the Environment Directorate to examine all of the issues and to debate them from various perspectives. In addition, officials of the Organisation have examined those papers of particular relevance to them. While many sources and experts have thus been consulted the best available knowledge and judgement on the various issues has been sought, the papers have been prepared by the Secretariat and do not necessarily reflect the views of OECD Member Governments.

The papers consider eleven issues. Of these, five concern environmental pollution issues: carbon dioxide and climatic change; the ozone layer; acid rain; chemicals; and the international movement of hazardous wastes. Resource issues are the subject of two papers: maintaining biological diversity, and loss of cropland and soil degradation. Four "management issues" are also considered: the environmental aspects of bilateral development co-operation; environmental impact assessment and international co-operation; environmental aspects of multinational investment; and the international application of the Polluter Pays Principle.

In developing these papers, an overriding theme emerged: the interrelationship between resource and environment issues on the one hand, and sustainable economic development on the other. This led to the preparation of the introductory chapter and to the title of this volume.

The papers focus on the international dimensions of the eleven selected issues, or on those aspects that could most benefit from co-operation at the international level. In this regard, a special report, included here as an Annex was prepared to provide an initial overview of the major activities relevant to these issues being undertaken by other international organizations. Although the organizations concerned have had an opportunity to review it, so that it should be up-to-date as of the end of 1981, it does not refer to all possibly relevant projects: it is selective and tries to cover the highlights.

One characteristic that marks the issues considered in this volume deserves special mention: it is uncertainty. Uncertainty surrounds certain issues perhaps more than others, issues such as carbon dioxide and climatic change, the ozone layer and acid rain. On these issues, governments are faced with a clear lack of consensus among scientists about their nature, their causes, or their consequences. This uncertainty is reflected in the papers where the reader will find ample use of the conditional tense.

Beyond reinforcing research to close the gaps of understanding, the question of how governments, industry and the public deal with uncertainty is an issue that itself needs further exploration. Society has learned, more or less, to deal with the uncertainty inherent in most economic and social issues. When uncertainty clouds understanding of possible changes in natural phenomena, however, it takes on a special significance, especially if those changes could have serious economic or social consequences. As one of the speakers in the Special Session pointed out, "In the absence of scientific consensus, political consensus, especially at the international level, is hard to achieve. However, if governments wait for scientific near-certainty, it will often be too late for them to act at all, either because of the inertia of natural phenomena or because of the time lags associated with policy development and international negotiation."(2)

Initially, these papers were addressed to the Environment Committee. The Committee felt, however, that they would be of interest to a wider audience and expressed the wish that they be published under the

---

(2) Ibid, page 11

responsibility of the Secretary-General. In reaching a wider audience it is hoped that they will serve to advance understanding and consideration of the issues.

Jim MacNeill
Director
OECD Environment Directorate

Paris 16th April, 1982

Introduction

ECONOMIC AND ECOLOGICAL INTERDEPENDENCE

Introduction

GLOBAL ECONOMIC AND ECOLOGICAL INTERDEPENDENCE

The growing interdependence of the international economic and political system has become a central question for governments and international organisations as they examine the critical issues likely to dominate the world scene to and beyond the turn of this decade and century. This interdependence is seen to cover not only population, migration, energy, food, financial transfers and technology, but also the environment and, increasingly, the ecological basis for development: renewable and non-renewable resources, the oceans, the atmosphere and climate, land, space and mankind's genetic resources and heritage. Understanding of this interdependence and its implications can lead to hope or despair depending on one's view of the will and capacity of governments to make necessary adjustments within and among their countries.

In this context, the interrelationships between population, resource and environment issues on the one hand and sustainable economic development on the other is receiving increased attention. A major theme of the US "Global 2000 Report" to the President, Japan's "Basic Directions in Coping With Global Environment Issues", OECD's "Interfutures" and IUCN's "World Conservation Strategy", it also features strongly in the Brandt Commission's "North-South: a Programme for Survival". It was the subject of a Special Session of OECD's Enviroment Committee in April, 1981(1) and it was also considered at the June meeting of OECD's Council at Ministerial Level. In July, the Ottawa Summit spoke of a "world of interdependence" and in its communique asserted that "in shaping our long-term economic

---

(1) See "<u>The Environment: Challenges for the '80s</u>" Proceedings of the Special Session of the OECD Environment Committee held on 1st April, 1981 concerning "OECD and Policies for the '80s to Address Long-term Environment Issues", Paris: OECD, 1981.

policies, care should be taken to preserve the environment and the resource base of the planet".

With greater understanding of the ecological basis for sustained economic growth, new approaches to strengthen multilateral co-operation will become more vital. Indeed a major implication of economic and ecological interdependence is that, as it inevitably increases, the ability of governments to deal unilaterally with problems on a national scale will diminish. More and more, economic, social, energy and other problems with an environmental or ecological basis within countries will prove resolvable or avoidable only through increased co-operation among countries.

While this is true of problems within and between OECD countries, it is also true of problems between OECD and developing countries. This is demonstrated by the OECD, US and Japanese reports cited above. Many of the global environment and resource issues considered in these reports result largely from the production or consumption patterns or the technology of OECD countries, and can be addressed through collective policy action on their part. Indeed, many of the resource issues that impinge most immediately on developing countries stem from the economic, trade and other policies of OECD countries. Moreover, the growing scale of issues such as the loss of genetic materials, the conversion of cropland, soil degradation and tropical forest destruction, and the inability of many Third World countries to deal with them, could have serious economic and even security consequences for OECD Members.

In this connection, it is useful to note that in 1976, investments by OECD countries such as Canada, Germany, Netherlands, Sweden, the U.K. and the U.S. in Third World countries represented about 25 per cent of their total foreign investments. In the case of Japan, the figure was 55 per cent. During the same year, imports from developing countries represented 28 per cent of total imports of OECD countries, while exports to developing countries represented 23 per cent of all exports. Around 29 per cent of all primary product imports by OECD, excluding fuel, originated in developing countries which, in turn, received 26 per cent of all OECD exports of manufactures.

These figures, interesting in themselves, add substance to the "Okita" Report's references to dependence "on overseas sources of energy, food, timber and other materials", and the need to "contribute actively to the conservation of the world's soils and forests, both for securing resources for its own sake and in order to prevent global deterioration of the environment".(2)

The Brandt Commission, too, was unequivocal in stating its view: "The world is now a fragile and interlocking system, whether for its people, its ecology or its resources.... More and more local problems can be solved only through international solutions - including the environment, energy and the co-ordination of economic activity, money and trade."[3]

Without closer international co-operation, increasing economic and ecological interdependence can also give rise to new and non-military threats to national security and survival. As the members of the Brandt Commission declared: "few threats to peace and survival of the human community are greater than those posed by the prospects of cumulative and irreversible degradation of the biosphere on which human life depends."[4] In the next section, their report continued: "Our survival depends not only on military balance, but on global co-operation to ensure a sustainable biological environment, and sustainable prosperity based on equitably shared resources."
The Brandt Commission then recommended that: "The world needs a more comprehensive understanding of security which would be less restricted to the purely military aspects."[5]

One of the speakers at the Special Session of the Environment Committee made exactly the same point, that the security of every nation depends on widening the definition of national security to include global environmental risks "that are virtually unknown to traditional diplomacy, that are beyond the reach of national governments, that cannot be fitted into received theories of competitive interstate behaviour, that are coming increasingly to dominate world affairs,

---

(2) Basic Directions in Coping with Global Environment Problems; Ad Hoc Group on Global Environment Problems, Government of Japan, December 1980.

(3) North-South: A Programme for Survival, London: Pan Books, 1980, p.33.

(4) op. cit.

(5) Ibid., pp.115-116.

that cannot be wished away, and that are indifferent to military force."(6)

The Need for Action

Present and prospective income levels and quality of life in developed and developing countries depend increasingly upon a global system with complex linkages not just between countries, but between various component parts of the system, whether they be political, economic or environmental. As illustrated by the above references and by the issues discussed in the volume, the destabilization of the world's ecosystem and the degradation of environmental systems in particular regions are among the fundamental problems which create actual or potential risks of concern to both developed and developing countries.

Speakers at the Special Session suggested that a 'global' approach, as distinguished from a "North-South" approach, would be a more appropriate description of the nature of international discussion and co-operation needed to enhance understanding among developed and developing countries on the nature of their mutual interests and individual responsibilities. They suggested that the time had come for the developed countries to formulate a more comprehensive, carefully considered and active approach to world economic security and to promote appropriate forms of international co-operation. Among other actions, they suggested that OECD should now try to bring into focus the various trends and factors which may significantly influence the effective and mutually beneficial functioning of the world economy into the future, including the special vulnerability of the developing countries.

Improving the Capacity to Anticipate and to Analyse Global Environment and Resource Issues

At the OECD Environment Committee Meeting at Ministerial level in May 1979, the Ministers adopted a "Declaration on Anticipatory Environmental Policies." Since that time, the reports cited earlier, the participants in the Special Session in April, and members of the Environment Committee have all stressed the need for an improved capacity within and among OECD countries not only to analyse global environment and resource issues but also, and more importantly, to put

---

(6) "The Extrapolation of Metaphors" by Harlan Cleveland in The Environment: Challenges for the '80s, Paris: OECD, 1981, p. 34 and p 48.

it all together, to examine the interdependencies and to relate the parts to the whole. Beyond this is the need to extract from such synthesis relevant conclusions that bear on policy action by Member governments.

At present, few OECD countries have such a capacity nationally, and some that do still face serious difficulties. For example, the "Global 2000 Report" found "serious inconsistencies in the methods and assumptions employed by the various agencies in making their projections.... It must be acknowledged that at present the Federal agencies are not always capable of providing projections of the quality needed for long-term policy decisions. While limited resources may be a contributing factor in some instances, the primary problem is lack of co-ordination."[7]

A recent Canadian study[8] makes essentially the same observations about the Canadian Government and, doubtless, examination would establish its validity for many other OECD Member countries as well.

At the international level, there is at present no mechanism through which governments may co-operate in the testing, review and comparison of the methods, results and implications of such studies.

OECD is well-placed to encourage, to facilitate, and to act as a catalyst for the development of an improved capacity on the part of Member governments to undertake co-ordinated and integrated analyses of environment and resource issues and their interdependence and other issues cited above and in recent major studies. For this reason, OECD has been requested to provide periodically a forum in which representatives of member governments and others undertaking such work could explore their ideas and problems and concert their efforts. Within this context, it could also monitor the results of major studies undertaken elsewhere, examine their interrelationships and, from time to time, extract relevant insights and conclusions that bear on policy action in various fields and draw them to the attention of Member governments.

---

(7) <u>The Global 2000 Report to the President on Global Resources, Environment and Population</u>, Washington: US Government Printing Office, 1980, Vol.1, pp.4-5.

(8) <u>Global 2000: Implications for Canada</u>: Toronto: Pergamon Press, 1981.

Selected Global Environment and Resource Issues

This report presents papers on eleven issues under three parts. Part One deals with "Environmental Pollution Issues" and addresses five problems (carbon dioxide and climatic change; depletion of the ozone layer; acid precipitation; chemicals; and the international movement of hazardous wastes) of growing international significance for which the causes and the capacity to avoid or to mitigate the consequences are to be found largely in OECD Member countries. Part Two deals with "Resource Issues" (genetic resources; and croplands and soil depletion) for which the principal causes and effects are still largely outside OECD Member countries but which, without strengthened international co-operation to understand and reduce the problems associated with them, could have serious consequences for agricultural production, medicine, the bio-technology and pharmaceutical industries and the economies generally of OECD Member countries. The third part addresses "Management Issues" and describes several areas and ways in which OECD Member countries could strengthen international co-operation and effort on selected global environment and resource issues.

The discussion of each issue concludes with a brief reference to the highlights of work under way at the international level and with some suggestions as to what else needs to be done. These suggestions are not exhaustive in nature and are designed to stimulate further thought on the issue.

During the preparation of these papers one of the experts involved observed that in the face of these issues, eventually requiring international as well as national responses, all countries are small. It was a dramatic yet simpler way of illustrating the theme of growing ecological and economic interdependence of all countries. Ten years ago Barbara Ward and René Dubos wrote a still relevant sourcebook under the inspired title of "Only One Earth". The official sub-title is more precise and down-to-earth, "The Care and Maintenance of a Small Planet". That sub-title, in a single phrase identifies the goal in the treatment of each of the issues described in the following papers.

Part One

ENVIRONMENTAL POLLUTION ISSUES

Chapter I

CO_2 AND CLIMATIC CHANCE

THE ISSUE

Increasing concentrations of CO_2 in the atmosphere, due mainly to the burning of fossil fuels, might cause a warming trend, leading to climatic changes in the next century of sufficient magnitude to produce major physical, economic and social dislocations on a global scale.[1] Significant uncertainties persist, however, as to causes and effects.

BACKGROUND

Carbon dioxide plays a critical role in maintaining the earth's heat balance by absorbing heat radiation from the earth's surface, trapping it, and preventing it from dissipating into space. Over the past century, CO_2 atmospheric concentrations are believed to have increased about 15 per cent largely as a result of the burning of fossil fuels and deforestation. Depending on growth rates in the burning of fossil fuels, global atmospheric CO_2 concentrations could double before the middle of the next century.

Such a doubling of CO_2 concentrations in the atmosphere could, according to estimates made using atmospheric modelling techniques, increase average annual global surface temperatures by about 2-3°C, and possibly as much as 7-10°C in the north polar region during the winter.

Temperature increases of this magnitude could, for example, produce changes in rainfall patterns,

---

(1) A team of NASA atmospheric scientists recently predicted that there is a high probability of warming in the 1980s and that the mean global temperature will increase by 1 to 4.5 degrees centrigrade by the year 2100. See Hansen, J., et.al., "Climate Impact of Increasing Atmospheric Carbon Dioxide", Science, 28 August 1981, Vol.213, No.4511, pp.957-966.

geographical shifts in areas suitable for food production and areas sensitive to desertification; higher sea levels due to the melting of polar ice; and changes in fish stocks, forests, and water supplies. Depending on their magnitude and timing, these changes could have profound social, economic and political impacts. Some changes might be beneficial to the region concerned, others deleterious, and differences in impacts, of course, could have repercussions on a global scale.

Increases in atmospheric CO_2 would also directly affect both agricultural crops and the biosphere. In general, one would expect photosynthetic efficiency to increase. However, plants react to increased CO_2 levels in many complex ways, and it is by no means certain that the net effect would be to increase productivity. In the biosphere, various species would react differently to increased CO_2 levels, thus altering ecological balances.

Clearly, the uncertainties at each stage of the chain of cause and effect from CO_2 emissions to societal impacts are very great. Neither the risk of any perceptible climate change due to CO_2 nor the possible impacts of such a change are well understood. However, if the potential long-term risks of increased atmospheric CO_2 should prove to be real, by the time the effects are perceived it may be too late to avoid appreciable climate change. It is therefore in the interests of all nations to seek to understand the phenomenon at an early enough stage to mitigate the risks of substantial economic, social and environmental disruptions.(2)

SIGNIFICANCE FOR OECD MEMBER COUNTRIES

OECD Member countries currently are responsible for about two-thirds of total world emissions of CO_2 by fossil fuel combustion. The contribution of the developing countries to emissions of CO_2 through fossil fuel burning will increase in the future but is likely to remain comparatively small. In this latter

---

(2) "Carbon Dioxide and Climate: A Scientific Assessment"; Report of an Ad Hoc Study Group on Carbon Dioxide and Climate, Woods Hole, Ma., July 23-27, 1979, to the Climate Research Board; Assembly of Mathematical and Physical Science, National Research Council (Climate Research Board, 1979; 35 pp; available from the board; supply limited).

group of countries the more pertinent issue is excessive deforestation and loss of humus in the soil.

The geographical distribution of the effects of climatic change would vary between and within regions and nations, but current models do not permit prediction of differential impacts with any confidence. They do, however, suggest that any temperature increases and therefore climatic change would be greater in higher latitudes than at the equator.

Any action to establish more precisely the nature of the risk and, if necessary, action to avoid or mitigate the future effects, would have to be initiated mainly by OECD Member countries.

CONSTRAINTS AND OPPORTUNITIES FOR ACTION

A major constraint may stem from the early lack of evidence of any significant temperature and climatic effects associated with an increase in atmospheric CO_2. Even if CO_2 emissions were to increase at a constant rate, temperature might not since the oceans can absorb both heat and CO_2 from the atmosphere up to a certain point. However, once the ocean's heat or CO_2 saturation point is reached, atmospheric temperatures would begin to rise at a much faster rate. Hence, a low rate of temperature increase in the near-term could mask the full extent of the problem in the longer run. In other words, evidence of climatic change may be both delayed and difficult to detect until CO_2 loading is such that appreciable climatic change becomes inevitable.(3) Consequently, action adequate to understand and anticipate such changes is necessary.

A major constraint is lack of adequate knowledge of the phenomenon. The work underway in national and international organisations on research, modelling and monitoring strategies should be strengthened. Additional efforts are needed to identify the possible geographical distribution of climate change. A limited research effort on the possible social and economic consequences is also required.

Although it may be too early to consider implementation of specific strategies, other than research and monitoring strategies, to respond to this issue, a number of strategies should begin to receive more attention as the results of research may warrant and make possible. These include:

---

(3) Ibid

- energy strategies, aimed at reducing the use of fossil fuels for certain purposes and at increased energy efficiency, a particularly favourable strategy for reducing pollution generally from all energy sources;

- development aid strategies, assessing aid assisted projects for their impact on the rate of deforestation and soil depletion or for the extent to which they actively encourage widespread afforestation; and

- adaptation strategies, to examine a number of "what if" scenarios and, as improved models provide a better picture of the probable geographical distribution and impact of climatic change, to use the results for advance planning of adjustment measures by nations and the international community.

OECD Member countries should give special consideration and attention in their bilateral aid programmes to measures to reduce deforestation and soil depletion, and to projects and measures to encourage afforestation. These problems deserve special attention in their own right, but deserve even greater emphasis due to their relevance to the issue of increasing atmospheric CO_2.

HIGHLIGHTS OF RELEVANT WORK AT INTERNATIONAL LEVEL

As will be seen from the Annex, several international organisations are involved in work directly related to this issue. The World Meteorogical Organisation (WMO) and the Environment Programme of the United Nations (UNEP), are active in climatic research and monitoring. In co-operation with the Food and Agriculture Organisation (FAO), the World Health Organisation (WHO), the United Nations Education, Scientific and Cultural Organisation (UNESCO), and the Scientific Committee on Problems of the Environment (SCOPE), they are also studying the carbon cycle, and they are assesing the impacts of CO_2-related climatic changes on agriculture, fisheries, water supply, sea level, health and natural ecosystems.

In February 1981, the OECD/IEA jointly convened an Experts Workshop on CO_2 Research and Assessment, in order to provide an initial review of the issue and of the type of work being undertaken to address it nationally and internationally.

SOME THOUGHTS ON WHAT ELSE MIGHT BE DONE

Nations should vigorously undertake and support efforts aimed at obtaining a better understanding of CO_2-related phenomena, its impact on global climate and the possible physical, social and economic changes associated therewith. In particular, work on research, modelling and monitoring needs to be strengthened and supplemented through increased international co-operation and joint efforts.

While OECD cannot itself undertake research, modelling or monitoring of the phenomena, it is maintaining a watching brief on the progress and results of scientific research and related activities at the national and international levels over the next few years.

In this regard, the OECD, in co-operation with the IEA and other international organisations, would intend to convene further meetings of experts periodically in order to assess progress and the state of the art; to review the trends and their implications; and to examine and provide guidance on the various strategies that may be required. Such meetings could consider the socio-economic implications of CO_2 build-up and reflect on the usefulness of further work in this area.

Chapter II

THE OZONE LAYER

THE ISSUE

The ozone layer plays an important role in protecting the earth from damaging ultraviolet radiation and research indicates that continued emissions of chlorofluorocarbons (particularly CFC-11 and CFC-12) and certain other substances could lead to depletion of the ozone layer. Significant uncertainties concerning these phenomena persist, however. Chlorofluorocarbons are used as propellants in aerosol cans, in foam production, and as solvents and refrigerants.

BACKGROUND

CFCs are not the only chemicals thought to affect the ozone layer. There is an array of man-made chlorinated compounds which are also thought to have an impact. The same is true of emissions of NO_x from aircraft and of nitrous oxide (N_2O) from the burning of fossil fuels and from the nitrification and denitrification of both organic and inorganic nitrogen fertilizers. Increasing concentrations of CO_2 in the atmosphere, through its effect on temperatures, could also affect the ozone layer by ameliorating to some extent the effects of ozone depleting substances.

One of the principal concerns about increased ultraviolet radiation resulting from any depletion of the ozone layer is an increase in skin cancer. Laboratory studies indicate that an increase in ultraviolet radiation might cause damaging effects on animal and plant life (certain crop species and marine organisms).

This document focusses on CFCs because of current concern about those chemicals and current responses to that concern. It is clear, however, that any research, modelling and control policies must consider all major chemicals which affect the ozone layer and this is reflected in the treatment of the subject matter in this document.

SIGNIFICANCE FOR OECD MEMBER COUNTRIES

In 1977, 90 per cent of the world production of CFCs took place in OECD countries. Eighty per cent of the CFC-producing facilities were located in OECD countries, and OECD countries were responsible for some 90 per cent of the world's consumption.

Production of CFC-11 and CFC-12 rose steadily from the 1930s until 1974 when it began to decline, a decline which continued through 1980, the latest year for which data are available. This continuing decline has been attributed to a reduction of CFC-11 and CFC-12 use in aerosols in response to changing economic conditions and consumer preferences, and to actions taken in individual countries by industry and government. It appears that the production of CFC-11 and CFC-12 manufactured by companies outside the OECD and the manufacture of other CFCs (not necessarily as significant as ozone depleters) has continued to grow, partially offsetting this decline.

Clearly, in the short term, concerted action could be taken by OECD countries to effectively control the level of emissions of CFCs world-wide, should that be viewed as necessary. In the longer term, however, there are wider international ramifications. It is already becoming apparent that other countries can move in to fill the vacuum created by a turn-down in production (and use) in OECD countries. Also, it must be remembered that CFCs offer major advantages over alternative chemicals in many applications, such as solvents and refrigeration, and that there is a need for the interests of non-OECD countries to be taken into account in any international policy considerations.

SOME POLICY OPTIONS

Reduction in emissions of CFCs can be achieved voluntarily or by imposing technical standards in the industries involved; by restricting or eliminating certain uses of CFCs; by limiting production or production capacity; or by a combination of these or other types of action.

Any action limiting CFC use would impose both long term and transitional costs on industry in terms of employment shifts, increasing production costs and changes in industry structure. The timing of the introduction of controls, and the length of time industry is given to adapt or adjust to any regulation, can have significant effects on the transitional costs (the longer the time given, the less the cost). The additional costs incurred over the long-term will not be affected by these factors unless it affects the type of control measures undertaken.

Thus, not only the measure, but the timing of its introduction in an individual OECD Member state or in the international community provides a series of options for consideration by governments and industry. In addition, the results of research on ozone depletion, ozone trend analysis and monitoring can be factored into plans to introduce any given measure and to the timing of its introduction, thus producing a further array of options.

CONSTRAINTS AND OPPORTUNITIES FOR ACTION

Since 1974 when the theory was first developed and predictions made that emissions of CFCs could deplete the ozone layer, there has been a great increase in understanding of atmospheric physics and chemistry. The subject has been studied using complex models to calculate variations in the depletion of the ozone layer with variations in projected emissions of CFCs and other chemicals. The results of the model calculations have changed over time as understanding of atmospheric physics and chemistry has improved.

Comparisons between model calculations and actual trends in ozone concentration are important for validation purposes. Data for the last 20 years, taking into account the accuracy of current monitoring techniques, show no evidence of a statistically significant change in total ozone. Current models have calculated a one per cent depletion due to CFCs alone and this depletion is not detectable with present techniques. The question of other perturbing emissions such as CO_x and NO_x has also to be taken into account.

Thus, a range of views exists concerning the degree of confidence which can be attached to calculations of ozone depletion, given present uncertainties in model formulation, chemical reaction rates, atmospheric and trend analysis parameters.

The uncertainties relate to: (i) long-term effects of emissions of CFCs and other man-made chlorinated compounds as well as CO_x, N_2O and NO_x; (ii) long-term effects of ozone depletion on health and the environment; and (iii) whether to take further control measures at this time, with attendant economic and social dislocations, as opposed to waiting for more research and monitoring data. As a result of these considerations a number of countries are reluctant to take action to halt the increase in CFC emissions at this time.

WHAT IS BEING DONE AT THE NATIONAL AND EUROPEAN COMMUNITIES LEVEL?

The European Communities, the United States, Sweden, Norway and Canada have taken actions to reduce the use of CFCs in aerosols. Additionally, the European Communities have taken action to ensure that their Member countries do not increase production capacity for CFC-11 and CFC-12. In Japan, the use of CFC's has been reduced as a result of voluntary efforts on the part of firms, and the Japanese Government is making efforts not to increase the production capacity of CFC-11 and CFC-12. Several other Member countries are considering plans to restrict further the production and use of CFCs, and the European Communities are re-examining the relevant scientific and economic information with a view to adopting such further measures as may be necessary.

Many countries not presently regulating CFCs are actively involved in the international discussion of the ozone depletion issue and in some cases voluntary actions have also been taken.

HIGHLIGHTS OF RELEVANT WORK AT INTERNATIONAL LEVEL

Several international organisations are involved in the ozone depletion issue. The United Nations Environment Programme (UNEP), through its Co-ordinating Committee on the Ozone Layer (CCOL), has been playing a major part in evaluating the CFC issue, particularly in the co-ordination and evaluation of research related to the ozone layer. The Governing Council of UNEP has also directed that work be initiated aimed at the elaboration of a proposed global framework convention on the protection of the ozone layer. This proposed action reflects the general concern about the need for international co-ordination in the assessment of the impact of man's activities on the ozone layer and responses to it.

The OECD, and particularly its Environment Committee, has also undertaken useful work on the ozone depletion issue. In 1976 it carried out an assessment of worldwide production, use and environmental impacts of fluorocarbons. More recently, another assessment has been made of the state of knowledge concerning effects associated with an increase in ultraviolet radiation, the state of development of modelling of the stratosphere and chemical impacts thereon, together with a review of the implications for certain industrial sectors of various possible control actions.

During 1981, this work was extended to a review by experts of various scenarios for continued emissions of CFCs with attendant global implications, in order to assist OECD countries to determine the need or otherwise for further control strategies at this time. The policy implications of this work will be examined in 1982 in relation to the need to protect the ozone layer.

Other organisations involved in the ozone depletion issue include the World Health Organization (WHO) (epidemiology studies), the World Meteorological Organization (WMO) (ozone monitoring) and the International Council of Scientific Unions (ICSU). Also, significant work is being undertaken by several industrial organisations, most notably by the Fluorocarbon Program Panel, an international group under the auspices of the US Chemical Manufacturers' Association.

SOME THOUGHTS ON WHAT ELSE MIGHT BE DONE

There is a major need to continue to improve knowledge of the chemical and physical processes of the stratosphere. This research should go hand-in-hand with refinement of existing one and two-dimensional models and improvement in techniques for monitoring and trend analysis. One of the particular challenges in understanding stratospheric ozone changes is the development of a variety of techniques to monitor ozone levels and trends at the various altitudes in addition to improving total ozone column measurements.

Existing appraisal of the economic and social impact of various policy options is impeded by a lack of knowledge of industry facts and figures, especially outside the United States. Options for control of emissions can be influenced by control technologies, substitute chemicals and production and usage patterns, thus calling for the development of a better data base in all these areas.

Work should continue on the effects of possible ozone changes associated with various emissions scenarios for both CFCs and other substances including associated health and environmental effects. Such work should also deal with the socio-economic implications of the scenarios. It would also have important implications for the international community in its continuing review of strategies to protect the ozone layer.

Chapter III

ACID PRECIPITATION

THE ISSUE

Sulphur and nitrogen oxides and other acid precursors emitted by natural and man-made sources can travel long distances in the atmosphere, undergoing chemical transformations, and returning to earth as acid precipitation. In sensitive areas, this increases the acidity of water bodies and the soil, and can damage aquatic ecosystems, crops and forests. Any satisfactory solution requires increased knowledge and a shared perception of the nature and extent of actual and future damage caused by acid precipitation and of the costs and benefits of counter-measures, so that appropriate strategies can be developed at both national and international levels.

BACKGROUND

Acid rain has been linked to emissions of sulphur and nitrogen oxides produced mostly by fossil fuel combustion in power plants, by smelting industries and from motor vehicle exhausts. Converted by chemical reactions in the atmosphere to sulphates and nitrates, the emissions return to the earth in rain or as dry deposition. Their effects are felt not only in the neighbourhood of the sources, but also at distances of hundreds of kilometres. Acid rain has been observed throughout the world and, particularly, in Scandinavia, in parts of the western U.S.A, in the north-eastern U.S.A and over large sections of eastern Canada. Much of it results from pollutants which have travelled distances of over 500 km.

It is now generally accepted that acid deposition can cause impairment of aquatic ecosystems (decrease in fish population of acidified lakes) and corrosion of materials. The potential effects on human health and on yield of forests and crops are under study. Evidence of the nature and extent of physical and socio-economic damage is accumulating. The effects occur both in countries which are major emitters of the gases, and in distant countries receiving the acid deposition as a result of prevailing wind patterns. Serious concern is being expressed in Scandinavia and Canada where the amount of damage from acid rain resulting from emissions in other countries may be significant.[1]

Projections of future emissions of SO_x and NO_x indicate a possible growth, notably if coal substitution for oil is carried out without adequate environmental control.

SIGNIFICANCE FOR OECD MEMBER COUNTRIES

OECD Member countries, because of their high industrial development over limited geographical areas, are large-scale producers of man-made SO_x and NO_x. In Europe, however, the contribution of Eastern European countries to acid rain may also be significant.

For these reasons work on acid deposition has been concentrated in industrialized countries, with international cooperation being channelled initially through OECD itself. The OECD "Programme on Long Range Transport of Air Pollutants" was commenced in 1972 and the results were published in 1977.

CONSTRAINTS AND OPPORTUNITIES FOR ACTION

While some of the consequences of acid precipitation are well established, the mechanisms underlying the chemistry of acid rain formation are complex and not completely understood. The transformation from oxides to acidic components is less well known with regard to NO_x than it is for SO_x. The role of oxidants and their precursors in the acid formation process remains to be clarified.

Nor is it clear how cost effective the available counter-measures would be in avoiding or reducing damage in recipient countries at long distances from the source. The relationship between costs and benefits is less well-known with regard to NO_x than it is for SO_x.

---

(1) See the recent "state-of-knowledge" report on this issue prepared by an international team of scientists: Atmosphere-Biosphere Interactions: Toward a Better Understanding of the Ecological Consequences of Fossil Fuel Combustion. National Academy Press, 2101 Constitution Avenue, Washington D.C., 1981.

Given the problems inherent in the methodologies, models and measures concerned, it will be difficult to establish the facts to a degree of precision sufficient to determine corrective measures satisfying all the parties concerned. Indeed, a major problem stems from the fact that the population in the source jurisdiction may have different priorities than that at long distance in the affected jurisdiction, and may place quite different values both on the costs of damage and on the costs of various control strategies.

Moreover, the current economic climate tends to reinforce the position of those who prefer to postpone investment in counter-measures, given that the ability to define and to cost the damage is far less developed than the ability to cost available technologies to control emissions.

Policy options for controlling the problem include the following:

(i) for energy production and consumption,

- more efficient use of energy;
- increased use of non-fossil fuels, low-sulphur fuels, and less polluting energy sources; and
- installation of the best commercially available technologies to reduce sulphur and nitrogen emissions, especially for new and significantly modified coal-fired power stations, which studies indicate can undertake such investment and remain competitive with oil-fuelled stations (though financing may be a problem for existing stations);

(ii) for smelting and transportation industries,

- R and D on new technologies to reduce emissions from the various sources of SO_x and NO_x and other possible precursors of acid deposition such as volatile organic compounds; and
- R and D on new process technologies.

SOME HIGHLIGHTS OF WORK AT INTERNATIONAL LEVEL

Considerable work relevant to acid deposition is now being carried out under international auspices. Several organisations within the United Nations system have programmes under way or planned, some involving other bodies such as the International Council for Scientific Unions. This is set out in the Annex.

As a follow up to the signing of the Convention on Long Range Transboundary Air Pollution (November, 1979) the U.N. Economic Commission for Europe (ECE) is developing a work programme on control technologies, on the effects of sulphur compounds, and on a review of national strategies and policies for the abatement of air pollution caused by sulphur compounds. The ECE is also active in the exchange of information on desulphurisation technology, having convened in 1981 the Third Seminar on Desulphurisation of Fuels and Combustion Gases.

In May 1980, the governments of Canada and the United States signed a Memorandum of Intent, calling for an investigation of the problem. The two countries are now engaged in a major research and monitoring programme in acid deposition, spanning the entire range from sources of acidic materials to impacts and techniques for emission control as well as deposition monitoring.

Previous work at OECD resulted in two major publications: "The OECD Programme on Long Range Transport of Air Pollutants - Measurements and Findings" (1977), and "The Costs and Benefits of Sulphur Oxide Control - A Methodological Study" 1981. Control technology for NO_x emissions in stationary sources is considered in a report to be published shortly. Work presently under way includes an evaluation of pollution control costs and their impact on coal's cometitiveness vis-a-vis other fuels. This is the subject of an international symposium to be held in the Netherlands in May 1982 under OECD patronage.

SOME THOUGHTS ON WHAT ELSE MIGHT BE DONE

Given growing concern about acid deposition in a number of regions and countries, and in the light of the constraints and opportunities for action discussed earlier, efforts to strengthen work at the national level and co-operation at the international level should be re-inforced. This applies, in particular, to efforts aimed at a better understanding of the science of the phenomenon, at improving control technologies and strategies and at assessing their cost-effectiveness.

International co-operation is needed in particular in the following areas:

(i) in further developing scientific understanding of the mechanisms and effects of acid deposition and of the economic and amenity implications of damage avoidance;

(ii) in determining the contribution of NO_x emissions and the role of oxidants and their precursors to acid deposition. A substantial amount of research has been devoted to sulphur emissions and the extent to which they are a cause of acid precipitation, but less research has been devoted to the role played by nitrogen oxides, and even less to that of oxidants;

(iii) in examining various control or remedial strategies and their associated costs; and

(iv) in evaluating the combined effects and costs of existing national strategies and various possible international strategies so as to indicate which are the most cost-effective and equitable strategies.

Chapter IV

CHEMICALS

THE ISSUE

For new chemicals, the issue is to develop and to introduce internationally harmonized schemes for their assessment prior to their entry into the market; while for chemicals which are already on the market, the issue is rather to establish criteria for their selection for further testing and assessment and for determining how best to harmonize and share the burden internationally.

BACKGROUND

As modern man's dependence on chemicals for societal benefits has grown, the potential for widespread pollution or contamination has also increased. There are now some 70,000 chemicals on the commercial market, and many of these are currently used and released into the environment with little or no knowledge of their potential long-range effects. It is estimated that 1,000 new chemicals enter the market each year.

Chemicals cross national boundaries in a number of ways. First they are traded in bulk and represent an important proportion of industrial production and trade of OECD Member countries. Second, chemicals reach virtually all parts of the globe in a vast number of internationally-traded articles including agricultural products, clothing, machinery, automobiles, etc. Third, they can be carried across national frontiers by wind and water.

The goal is to develop management strategies which, on the one hand, reduce the risk of unintended effects of chemicals on human health and the environment, but on the other hand, ensure the continued vigour and innovative capacity of the chemicals industry.

SIGNIFICANCE FOR OECD MEMBER COUNTRIES

The chemicals industry plays a critical role in the economies of OECD countries, with more than $450 billion in annual sales in 1980, $120 billion in exports and four million jobs involved. There are few industries which are not served by chemicals and examples of dependence for health care, agriculture and consumer product manufacture on chemicals in OECD's industrialized societies are legion.

Consumption of chemicals in OECD countries amounts to more than $400 billion, and more than 80 per cent of the overall world trade in chemicals takes place among OECD Member countries. This means that a high percentage of the problems in production and use of chemicals are first recognised or anticipated and forestalled in OECD countries. Increasing information on potential long-term chronic effects of some chemicals is revealing the significance of the management problem for chemical-dependent societies. The question can be posed whether the rate of growth of development of understanding and of institutions to deal with the issue is sufficient to address the problem.

OECD countries have a major role and responsibility as the primary producers and users of chemicals. Progress in managing contamination of the global commons will depend on the ability of OECD Member countries to develop effective policies concerning the production, use and disposal of chemicals and in so doing to anticipate their potential hazards. Central to effective management of chemicals will be the availability of consistent and adequate information to all countries in order to enable each to take informed decisions.

THE EVOLUTION OF POLICY DEVELOPMENT IN OECD

It is widely recognized that the case by case approach which reacts to identified hazards of specific chemicals does not provide an adequate basis for developing anticipatory management policies for chemicals, at either the national or international level.

Since 1969, governments of many OECD countries have moved to develop or to enact pre-market or pre-manufacture chemical control legislation to ensure the protection of human health and the environment, while at the same time trying to reduce any adverse impact of such legislation on economic activity and trade. To date 10 of 24 OECD governments have enacted such legislation.

At the international level the challenge has already been accepted within the OECD to promote the harmonization of these national laws, approaches and procedures through an active effort to develop national policies in the light of internationally agreed goals.

CONSTRAINTS AND OPPORTUNITIES FOR ACTION

Legislation already enacted by a number of Member countries gives clear indication of the scope which exists for policy action. But, at the national level there is the difficult task of integrating control strategies across an array of different types of chemicals, some of which have been regulated for many years, e.g. pharmaceuticals and pesticides, and others which, like industrial chemicals, have only recently attracted attention in most OECD countries.

At both the national and the international level it is difficult to develop comprehensive strategies which provide a framework within which both new and existing chemicals can be controlled.

A significant obstacle to the effective control and assessment of existing chemicals lies in the lack of knowledge of, and access to data on, any given marketed chemical especially in cases of imported chemicals. The costs of developing data and the confidentiality of some data are further constraining factors.

WHAT IS BEING DONE AT THE NATIONAL LEVEL

The move towards the enactment of anticipatory legislation has been gaining momentum in OECD countries. The first to enact such legislation was Switzerland in 1969 where every chemical now entering the market is classified according to its potential danger to human health. In 1973, Japan and Sweden, as leaders in environmental matters, passed laws aimed at broadened chemical controls which included environmental considerations. The United Kingdom, Canada and Norway followed between 1974 and 1975. This trend was given a strong boost in 1976 by the passage of legislation in the United States, one of the world's major chemical producers. By early 1977, France had passed similar legislation and Denmark and Germany followed suit in 1979 and 1980 respectively, after the introduction in the European Communities of a major Directive. This Directive requires legislation to be introduced in Community countries and provides the basis in law for systematic scrutiny of the potential health and environmental hazards of chemicals. New Zealand launched its Toxic Substances Act in 1979.

In addition, legislative actions are currently pending in the Netherlands, the United Kingdom and Switzerland. These actions provide for general health and environmental controls on chemicals through enactment of new laws or by complementing existing ones. Other OECD countries have legislation in draft form.

HIGHLIGHTS OF RELEVANT WORK AT INTERNATIONAL LEVEL

By the early 1970s international attention was focussing on those chemicals which had not formerly been regulated. Consideration of these chemicals, which were largely industrial chemicals, benefited from the earlier work of the Food and Agricultural Organisation (FAO) on pesticides, and of the World Health Organisation (WHO) on pharmaceuticals and food additives.

Since that time significant work has been undertaken by the UN and the OECD. Commencing with the UN Conference on the Human Environment in 1972, the following programmes have been established within the UN system: the International Register of Potentially Toxic Chemicals (UNEP), the International Occupational Safety and the Health Hazard Alert System (ILO), and the International Programme on Chemical Safety (WHO, ILO, UNEP).

The work of the OECD has been directed mainly towards the development of harmonized approaches to the management of new chemicals. Much of this work has established principles and approaches which are also relevant to existing chemicals.

In 1980, the OECD High Level Meeting on Chemicals called for major international actions to ensure harmonization of national chemical control policies. In 1981, the OECD Council took up these proposals and adopted a Decision on the Mutual Acceptance of Data. This Decision was supported by two Recommendations; one on Test Guidelines to be used in anticipating the effects of chemicals; and another on Principles of Good Laboratory Practice to assure the development of high quality data. At the time of writing, a draft Decision on a proposed Minimum Pre-marketing set of Data to be used in the initial assessment of a chemical prior to marketing is before the OECD Council.

The long-term global issues associated with chemicals show the need and urgency for internationally harmonized strategies to control existing as well as new chemicals and the need to consider the importing as well as the producing countries. This was recognized by the Council in 1981, based on advice from its Committees, when it supported the continuation and development of a comprehensive future programme with broad thrusts aimed at:

(i) assessment of new and existing chemicals and development of harmonized and effective legislative controls;

(ii) the updating and implementation of recent and pending Council actions;

(iii) the analysis of potential economic and trade impacts of chemicals control; and

(iv) the development of mechanisms to facilitate information exchange on chemicals.

SOME THOUGHTS ON WHAT ELSE MIGHT BE DONE

Major efforts are required on the part of individual OECD Member countries to implement quickly those measures which have been adopted by the OECD Council.

In order to encourage the wider use of technical instruments such as the Test Guidelines and Principles of Good Laboratory Practice co-operative arrangements need to be fostered between OECD and non-OECD countries and strengthened between OECD and other international organisations. Such arrangements as have evolved to date reveal that there are significant mutual advantages to be gained.

There is also the continuing challenge of ensuring that policies and agreements are more widely applicable than within OECD alone. OECD's work on chemicals already takes account of many global issues and needs. In reviewing the further direction of OECD's work, however, experts have raised many interrelated questions focused on the need for an international framework to integrate chemicals activities, both conceptually and operationally. They pointed to the scale of global use of chemicals; the impact of chemicals on the pattern of life in all societies; the need for environmentally beneficial alternative chemicals and technology; the interlinkages between policies for management of chemicals in the work-place, chemicals in transport, testing and assessment of chemicals and ultimately their disposal.

Chapter V

INTERNATIONAL MOVEMENT OF HAZARDOUS WASTE

THE ISSUE

The management of hazardous wastes requires an appropriate framework of policies both for the export and the import of such wastes between countries.

BACKGROUND

During recent years thousands of abandoned storage sites for hazardous wastes have been located in North America and in Europe. Events such as those at Love Canal (United States) and Lekkerkerk (Netherlands) have led governments of OECD Member countries to consider new and more effective measures to deal with the registration, movement, treatment and safe long-term disposal and storage of hazardous wastes. These measures normally require the proper siting of treatment and storage facilities, a process that is proving difficult in some Member countries because of public concern and opposition.

The "not in my backyard" syndrome and the growing cost of treatment, storage and disposal of hazardous wastes provide strong incentives to "export" such wastes to other regions or countries for treatment or for temporary or permanent storage.

There are various legitimate reasons for such exports: for example, treatment facilities in a neighbouring country may be closer to the source of the wastes than similar national facilities; a country may not have the capacity to treat certain wastes within its own boundaries; disposal facilities for some specific wastes exist only in very few places (as in the case of underground disposal); or waste material generated in one country might be used as a raw material input in industrial processes in another country (leading to a waste exchange market).

At the same time, however, under other conditions, the export of hazardous wastes may simply reflect a search for a jurisdiction in which environmental awareness and regulations are weak or non-existent.

SIGNIFICANCE FOR OECD MEMBER COUNTRIES

Most (approximately 80 per cent) of the world's hazardous waste has accumulated, and is currently produced, in OECD Member countries. The need for comprehensive policies to manage hazardous wastes has therefore arisen in OECD countries, and the benefits of sound management will accrue mostly to them.

Because of the volume of waste generated by OECD countries, they have a particular interest and responsibility to develop a sound policy framework for the export and import of hazardous wastes. The controlled movement of such wastes between countries may provide benefits for all. The uncontrolled movement of such wastes, however, could result in serious long-term damage to health and property in importing countries and to the global commons.

CONSTRAINTS AND OPPORTUNITIES FOR ACTION

There are serious practical difficulties in regulating the export or import of hazardous wastes.

The main regulatory responsibility should normally rest with the importing country, and this would be effective providing the country concerned possessed the institutional infrastructure, laws and knowledge to deal properly with imported wastes.

The importing country, however, may have no knowledge of the credentials of the exporting company or agency, nor of the nature and properties of the waste concerned. It might request this information from the exporting country. It might also request authorities in the exporting country to certify that any information given by the exporting company or agency is complete and accurate. In such cases, this would imply some regulatory mechanism on the part of the exporting country.

In certain circumstances, private and public bodies in exporting countries may find it comparatively advantageous to export their waste with minimal curiosity and regulation on the part of the importing country. Moreover, they could be aided in this by the lack of an institutional structure, or adequate laws, knowledge or experience concerning wastes on the part of many importing countries, especially outside the OECD.

Most countries, in fact, do not yet have institutions, laws or policies to control waste exports and imports. There is ample opportunity, therefore, for action and international co-operation aimed at ensuring that imported and exported wastes are disposed of or treated in an environmentally safe and cost effective manner.

HIGHLIGHTS OF RELEVANT WORK AT INTERNATIONAL LEVEL

As may be seen in the Annex, several international organisations within the United Nations system have become involved in the issue of hazardous wastes, including UNEP, WHO, ECE and the United Nations Development Programme (UNDP).

In April 1980, the UNEP Governing Council called for the development, in co-operation with competent authorities, of guidelines for the safe and appropriate disposal of hazardous chemical wastes and pertinent measures concerning their transboundary transport; and draft guidelines are being developed. In co-operation with UNEP, the WHO in 1981 prepared a draft code of practice for the management of hazardous waste.

In November 1981, an Ad Hoc meeting of senior government officials expert in environmental law recommended that "guidelines, principles or agreements should be developed on the transport, handling and disposal of toxic and dangerous waste". This recommendation is to be submitted to the UNEP Governing Council in May 1982.

The UN Economic Commission for Europe is considering measures for the safe handling and control of toxic wastes, including modified technologies or recycling techniques to reduce waste generation, and for the issuence of annual reports on activities of ECE governments and international organisations on the control of toxic wastes.

In 1978, the Commission of the European Communities adopted a Directive on Toxic and Dangerous Waste. A draft directive on the transfrontier transport of hazardous waste is now in preparation.

In 1976, the OECD Council, acting on proposal of the Environment Committee, adopted a Recommendation on Principles Concerning a Comprehensive Waste Management Policy. Embracing the cycle of design, manufacture and use of products as well as the reclamation and disposal of waste, it aimed at the most efficient and economic reduction of nuisances and costs generated by wastes. Thus it stressed reduction at source, reclamation and recycling as well as economic policy instruments reflecting the Polluter Pays Principle.

During the past two years, the OECD has had an active programme on the management of hazardous wastes. Currently, OECD's programme is focused on (i) the transport of waste across national frontiers, the quantities transported and the reasons for such transport; (ii) the costs to industries of complying with national hazardous waste management regulations, and the costs to the regulating agency of enforcing regulations; and (iii) the financial provisions necessary to ensure the appropriate handling, treatment and disposal of hazardous waste, and the liability for damage caused by inappropriate management at any stage. Reports on these aspects of the hazardous waste issue will be published by OECD.

SOME THOUGHTS ON WHAT ELSE MIGHT BE DONE

In view of the growing importance of this issue and its international implications, an important, though not exclusive, focus for international co-operation is the development of guidelines for the export and import of hazardous waste. A first exchange of views on waste exports problems took place in the OECD Waste Management Policy Group in April and November 1981. Further work is under way and will continue in 1982.

Any guidelines developed would necessarily relate to the export and import of wastes by OECD Member countries and would presumably reflect the differing levels of development of environmental regulations and the different levels of expertise among them. They should also reflect the economic dimension of the problem, seeking cost-effective means to regulate wastes, while at the same time ensuring that the export of hazardous wastes does not become a source of distortion of trade.

Within this broad area, work could be directed in particular towards the development of certain common measures for tracing and regulating flows of hazardous wastes between countries of similar infrastructure, knowledge and experience, and the development of a set of guidelines to deal with exports to other countries, including any necessary assistance to those countries to establish the appropriate institutional infrastructure, laws and experience to deal with imported waste.

Part Two

RESOURCES ISSUES

Chapter VI

MAINTAINING BIOLOGICAL DIVERSITY[1]

THE ISSUE

The genetic base of many of the world's crops and livestock has narrowed, making them more vulnerable to pests and diseases and to changes in soils and climate. At the same time, the world's genetic resources-essential for reducing that vulnerability, as well as for producing a great range of pharmaceutical and industrial products - have been depleted, and this trend continues. Some 25,000 plant species and more than 1,000 vertebrate species and sub-species (2) are now known to be threatened with extinction, and as much as 10 per cent or more of all species on earth could be extinguished over the next two decades. Extinction of species on this scale is without precedent in human history.

The most serious threat to species is the destruction of their habitats, such as wetlands and forests. Tropical forests, in particular, harbour over one-third of all species world-wide. In many developing countries the tropical forests are being exploited at a rate which scientists consider ecologically and economically unsustainable. A rapid decline of tropical forests and their genetic resources over the next two decades is likely to have serious environmental, economic and social consequences for developing and developed countries alike. Yet the magnitude of the problem and the likely consequences, and especially of the implications for present national development and

---

(1) The term "biological diversity" embraces both genetic diversity (the variability in a given species) and ecological diversity (the number of species in a community).

(2) These figures are given in The World Conservation Strategy (Chapter 3) and are based on the IUCN Red Data books. The Global 2000 Report estimates that 15 to 20 per cent of all species on earth could be extinguished by the year 2000 (see Volume One, p.37).

aid policies, have only begun to be assessed. Furthermore, the loss of wetlands and forest habitats is endangering foodchains and overall ecological stability, so the problem is more than just a threat to certain genetic resources.

The destruction of species and genetic materials has been and is opposed on ethical grounds, but their protection is also a matter of self-interest for all countries.

BACKGROUND

Genetic materials and biological diversity are foundation stones for global economic development, food security and the supply of fibres and certain drugs. This is particularly true for OECD Member countries.

The breeding of plants for high yields, improved quality and greater resistance to disease and pests has been one of the most spectacular successes of agricultural research. More than 70 per cent of the crop production in the USA, for example, is based on plant species brought in from outside, and, as in other OECD countries, almost all crops contain genetic material from a number of developed and developing countries.

Many major crops in OECD Member countries have a restricted genetic base. For example, only four varieties of wheat produce 75 per cent of the crop grown on the Canadian prairies; and more than half the prairie wheatlands are devoted to a single variety (to Neepawa, derived in part from germplasm supplied by Kenya in the late 1950's). Similarly, 72 per cent of US potato production depends on only four varieties; just two varieties supply US pea production; and the US soybean industry had its early origins in just six plants from one place in Asia, although since then it has benefited from genetic material from other countries.

As genetic diversity is reduced, however, so is the capacity of scientists to develop new and improved commercial species; and the genetic base of the world's present and future crops and other living resources is being rapidly reduced. Many wild and domesticated varieties of food plants - such as wheat, rice, millet, beans, yams, tomatoes, potatoes, bananas, limes and oranges - are already extinct, and many more are in danger. Useful breeds of livestock are also at risk: 115 of the 145 indigenous cattle breeds in Europe and the Mediterranean region are threatened with extinction.

Medicines and other pharmaceutical products are heavily dependent on plant and animal species. It is estimated, for example, that more than 40 per cent of the prescriptions written each year in the USA contain a drug of natural origin - either from higher plants (25 per cent), microbes (13 per cent), or animals (3 per cent) - as the sole active ingredient or as one of the main ones. It is also noted that the commercial value of all medical preparations in the USA derived from natural origins now surpasses $10 billion a year. The most important applications of higher plants and animals in the bio-medical area are as constituents used directly as therapeutic agents; as starting materials for drug synthesis; and as models for drug synthesis, for toxicity testing and for serum preparation.

Another area where OECD countries are vitally dependent on species now threatened with extinction is bio-medical research. A wide range of bio-medical programmes, both public and commercial, make extensive use of exotic animals from developing countries. An important part of these specialised programmes cannot be carried out on captive-bred specimens, or its needs cannot be met by the limited number of specimens available through captive breeding. Much research hence depends on a continuous supply of wild-caught animals. For example, African chimpanzees, now on the endangered species list, were considered to be "an irreplaceable surrogate for man" in biochemical research [3], especially for the development of vaccines against infectious diseases such as hepatitis.

Many other primates, including all of the great apes, are now thought to be endangered. This is due to a combination of habitat destruction (yet another aspect of the tropical forest syndrome); commercial demand from developed countries, not only to meet bio-chemical research needs, but also for zoological collections and pet trades; and, in some cases, local hunting for food. Less than 20 per cent of the primates currently used in bio-medical research are produced by captive-breeding. Furthermore, the success of major breeding programmes, existing or planned, as potential suppliers for the research needs of OECD countries will depend primarily on the conservation of suitable habitats for the species concerned.

---

(3) U.S. Interagency Primate Steering Committee, "Report of the Task Force on the Use and Need for Chimpanzees", National Institute of Health (Bethesda, Maryland 1978) p.3

In addition to containing about one-third of the world's plant and animal species, tropical forests provide a wide variety of goods, useful to affluent industrial and poor rural communities alike, including timber, pulpwood, fuelwood, fodder, fruit, fibres, pharmaceuticals, resins, gums, waxes and oils. Many developing countries are heavily dependent on their tropical forests; eight of them each earn more than $100 million a year from exports of forest products.

Tropical forests, however, are being exploited to-day at an unsustainable rate, with little regard for their global, regional or even local ecological value. They are contracting rapidly as a result of expanding and shifting agriculture, spontaneous settlement, planned colonisation, clearance for plantations and ranching, cutting for fuel and logging.(4) In order to protect their usefulness as habitats, scientists estimate that at least 10 per cent, and possibly 20 per cent, of the total area of tropical forests needs to be set aside as protected areas. At present, only two per cent of such forests are protected, and many developing countries lack the institutional and financial capacity to protect even this proportion.

As yet, only about ten per cent of the world's plant species and one per cent of the world's animal species have been subjected to a preliminary screening to establish their medical and other commercial applications. Bio-technology is expanding rapidly while, at the same time, the genetic resources on which its innovative capacity and future depend are declining.

At present, it cannot be predicted what other species may become useful to man. Many species that may

---

(4) The Global 2000 report states that because so little is known about the process of deforestation and the rates at which it occurs in many different countries, it is "impossible to predict the global conditions of forests in the LDCs in the year 2000". However, later in the report is "a mildly optimistic scenario" which suggests that by the year 2000 there could be losses of between 20 per cent (in Africa) to 50 per cent (in Asia and the Pacific). Overall, the Global 2000 Report states that "projections indicate that by 2000 some 40 per cent of the remaining forest cover in LDCs will be gone." (Volume One, p.2). More recent analyses by FAO and others suggests that the loss may be smaller than this estimate but it is nevertheless a substantial and unsustainable rate of exploitation.

seem to be of little consequence could in fact have important medical and commercial applications, or be key elements of the major life-support systems and processes on which mankind depends. The accelerating loss of genetic resources is an issue which needs to be addressed now for reasons of ethics for some, and of self-interest for all.

SIGNIFICANCE FOR OECD MEMBER COUNTRIES

Over the last three decades, about one-third of the gain in agricultural productivity in OECD countries resulted from animal and plant breeding, with much of the germplasm originating from tropical and sub-tropical areas. It is estimated, for example, that US wheat production has been boosted by about $500 million a year by imported germplasm.

For OECD countries as a whole, it is estimated that 10 plant species account for 70 per cent or more of total cash receipts from crops. Future increases in productivity and decreased vulnerability to new pests and diseases, and to changes in soils or climate, will depend substantially on genetic resources originating outside OECD Member countries.

A preliminary inventory of 50 countries carried out by IUCN for the International Board for Plant Genetic Resources (IBPGR) in 1980-81 showed that among the wild relatives of crop plant species and subspecies, at least 25 are now known to be threatened by extinction and urgently require protection. Most of these are not located in protected areas. Indeed, few nature protection areas maintain adequate data on the crop genetic resources they contain.[5]

CONSTRAINTS AND OPPORTUNITIES FOR ACTION

In spite of the fact that the Stockholm Action Plan of 1972 gave special attention to genetic resources and the need to preserve them as a matter of priority, they are not yet viewed as a strategic global resource. Consequently, the economic, social and political implications of rapidly declining genetic resources have only begun to be assessed.

---

(5) Robert and Christine Prescott-Allen, "In Situ Conservation of Crop Genetic Resources", International Union for Conservation of Nature and Natural Resources (Gland 1981); abridged version to be published in "Nature and Resources" (UNESCO, Paris).

About seventy per cent of the world's genetic resources are located in developing countries, which themselves have little capacity to protect or to exploit them and, for the sake of their own economic development, tend to encourage activities such as the exploitation of forests, which pose the greatest threats to these resources. At the same time, genetic resources, especially in developing countries, are dealt with as "free-goods" and the host country derives little or no economic benefit from their exploitation. The scientific and technical expertise to exploit species' genetic resources is confined largely to OECD and other developed countries.

It is difficult for most developing countries simply to set aside any sizeable territory in order to preserve forest ecosystems. The short-term opportunity costs can be substantial. In addition there are the costs of actually managing the protection of designated areas. For many developing countries, and the rural poor which constitute the majority of their population, there is simply no choice. Without alternatives, or assistance with management, their short-term requirements for economic development, for fuelwood and for physical survival takes precedence over the need to preserve the tropical forest and its genetic resources.

Environmental policies in OECD countries may also be contributing to the destruction of the tropical forest. In his report to the Japanese Government, Mr. Okita observed that "an action that may be reasonable in terms of protecting one country's environment may not be so in terms of environmental protection on a global scale." The example he cited was the "felling of tropical forests overseas because forests in one's own countries are protected."(6) The example also serves to highlight the complexity of economic and ecological interdependence on a global scale, in that the extension of protected parklands in an OECD Member country might ultimately compromise future agricultural production in that country by contributing to the loss of tropical forests and biological diversity in some distant country.

Several international programmes to maintain the world's stock of species are managed by FAO, UNESCO and other bodies. Their objectives - such as the UNESCO effort to establish a global network of biosphere reserves representing the world's 200 biotic provinces -

---

(6) Basic Directions for Coping with Global Environmental Problems, Report to the Government of Japan, December 1980, p.11

are far in excess of the financial resources available to serve them. Given the growing scale of the problem, these programmes should be strengthened. Major efforts to promote species conservation have also been undertaken by IUCN and the World Wildlife Fund whose resources have come largely from non-governmental sources. These programmes, too, merit additional support. The same is true of species conservation programmes at the national level such as those maintained by the U.S.A., Netherlands and Germany.

There are many possibilities for action at the national and international level. Priority measures include intensive surveys to identify species and habitats threatened with imminent extinction; increasing the number and range of protected habitats of international significance; establishing special programmes of financial and technical assistance to help developing countries protect unique habitats and species; developing alternatives to the economic activities which threaten them; and expanding existing national and international germplasm banks to cover a wider range of plant and animal species.

In the case of tropical forests, the World Conservation Strategy includes many recommendations for national and international action: the establishment of fuelwood and industrial plantations in critical areas; reforestation of strategically located forests; the strengthening of institutions and management of forests and protected areas; the establishment of networks of protected areas to safeguard a comprehensive range of the genetic diversity of tropical forests; and the development of systems of commercial exploitation that utilise products other than timber (for example, drugs, gums and resins).

It has also been suggested that genetic resources are, and should be declared to be part of the world's common heritage, to be maintained on behalf of the global community. To this end suggestions have been made for a comprehensive convention regarding the preservation of genetic resources on site and in gene banks.

There appear to be few alternatives to providing further development assistance to the developing countries in which most genetic resources are located, but the possibilities for an international levy on, or marketing of, genetic resources could be explored.

HIGHLIGHTS OF RELEVANT WORK AT INTERNATIONAL LEVEL

Certain references have been made in the above to activities being undertaken or proposed at international level, and it will be seen from the Annex that several international organisations have been and are involved in work directly related to the maintenance of biological diversity.

The Convention on International Trade in Endangered Species of Wild Fauna and Flora is an important instrument in this regard. Adopted in Washington in 1973, the Convention entered into force in 1975 and there are now 82 signatories. The International Union for the Conservation of Nature (IUCN) with the United Nations Environment Programme (UNEP), the Food and Agriculture Organisation (FAO) and the World Wildlife Fund (WWF),work together in encouraging governments to ratify this and other international conservation treaties and to assist them in formulating appropriate national legislation.

The World Conservation Strategy, mentioned earlier, is another vital instrument. Prepared by the IUCN, with the support of UNEP and the WWF and with the co-operation of FAO and UNESCO, it aims to maintain essential ecological processes and life-support systems; to preserve genetic diversity; and to ensure the sustainable utilization of species and ecosystems. During 1982-83, these same bodies propose to conduct systematic monitoring of, and to issue assessment statements on, the status of certain plant and animal species. They also propose to prepare action plans for the conservation of wild genetic resources.

The FAO, in co-operation with several organisations, supports the collection of specified germplasms for crop plants according to approved priority lists, evaluates and stores them, and publishes information on them and on methodologies for their on-site conservation.

The ECE, in co-operating with the Council of Europe and UNESCO, is currently evaluating international measures taken or being considered for the conservation of flora and fauna, and their habitats in the ECE region. It is also examining the possibility of a network of representative ecological areas in the ECE region.

The Annex reveals that there is also significant work under way or planned concerning fish and forest resources.

SOME THOUGHTS ON WHAT ELSE MIGHT BE DONE

Given the foregoing analysis, it is clearly important that on-going work to develop and implement conservation strategies be re-inforced and strengthened at the national and international levels. The same is true of the work to monitor, to collect, and to assess the potential uses of plant and animal species; to develop plans for on-site conservation; and to extend the network of representative ecological areas and the other activities mentioned in the Annex.

There is, however, one dimension of maintaining biological diversity, that appears to have been largely neglected to date: the economic dimension. This is due, probably, both to inadequate methods of assessment and to sheer lack of information and data. Indeed, it would appear that little or no work has been done to lay the foundation of information that would permit estimates of the potential economic value of genetic materials to be made. Work in this area could be of considerable benefit and OECD is giving some thought to it.

Such work might be undertaken on a case-by-case basis to determine the extent of certain uses of genetic materials by selected industries, including the agricultural and pharmaceutical industries. It could lead to examination of the economic benefits of applications to date and of future prospects, including the costs and possible environmental consequences of synthetic or natural substitutes. It might also enable some useful conclusions to be drawn on methods of evaluation.

Chapter VII

LOSS OF CROPLAND AND SOIL DEGRADATION

THE ISSUE

Most of the land in the world that is best suited for crop production is already being farmed. If the current rates of conversion of agricultural land to non-agricultural uses continue in OECD Member countries, and if the current rates of land degradation continue in developing countries, within 20 years more than one-third of the world's arable land could be lost or destroyed.

BACKGROUND (1)

Only about 11 per cent of the world's land area (excluding Antarctica) offers no serious limitation to agriculture. The rest suffers from drought, mineral stress (nutritional deficiencies or toxicities), shallow soil depth, excessive water or permafrost. The world's cropland currently occupies 14 million km^2. and, even using the most optimistic assumptions, it would appear that croplands world-wide could be no more than doubled.

Within OECD Member countries as a whole, large areas of prime quality land have already been lost permanently for agricultural purposes, primarily through urban and industrial development, recreation development and the construction of transportation infrastructure and reservoirs. At least 5,000 km^2 of cropland in OECD Member countries are lost each year to urban

---

(1) Much of the information in this section, unless otherwise indicated, is derived from the World Conservation Strategy issued last year by the IUCN, UNEP and the WWF, especially Chapter 2 on the "Maintenance of essential ecological processes and life-support systems", which is itself based on the most recent FAO and UNEP studies.

sprawl.[2] In the United States alone, about 2,500 km^2 of existing cropland and 700 km^2 of potential cropland, are converted each year to urban and built-up uses.[3]

There are, unsurprisingly, differences among OECD Member countries in the kind and magnitude of changes in land use. Over the last twenty years in nearly all European countries and in Japan, the arable land area has shrunk, while in less populated countries such as Australia and New Zealand, it has remained stable or increased. But even in countries where total arable land is growing, the area of prime agricultural land may still be getting smaller, as is the case in Canada.

Currently such losses are compensated by intensifying production on the remaining land or developing new cropland from existing pastures, forests or wetlands. These measures have a number of undesirable environmental consequences, however. Drainage of wetlands destroys habitats and threatens species and food chains as noted in the previous chapter. Much of the new land is more marginal than that which it replaces; hence more fertilizer is required, soil erosion is greater, and loss of nutrients and pesticides to surface and groundwater increases.

Erosion is already a serious problem in some OECD Member countries. For example, in the USA and Australia as much as 50 per cent of the cropland is already degraded by wind and water erosion. In the USA, one-third of the cropland is losing topsoil faster than the rate of natural soil generation. For all practical purposes, soil is a non-renewable resource: once it is gone, the loss is permanent. Nature takes from 100 to 400 years, or more, to generate 10 millimetres of top soil, and 3,000 to 12,000 years would be needed to generate soil to a depth equal to the length of the page on which this sentence is printed.

Developing countries are also experiencing an increasing loss of arable land, in part because of conversion from agricultural to non-agricultural uses, but largely because of soil degradation. Soil loss has accelerated sharply in most developing countries, which

---

(2) See The State of the Environment in OECD Member Countries, Paris: OECD, 1979, especially Chapter 2 on "Land", pp.65-75.

(3) Final Report of the National Agricultural Lands Study, Washington: US Government Printing Office, 1981, p.2.

are generally more susceptible to erosion than most OECD Member countries due to the topography of their lands, the nature of their soils and their climate. More than half of India, for example, suffers from some form of soil degradation. Out of the total of 3.3 million km^2 of the Indian sub-continent, over 40 per cent of the land is subject to increased soil loss and an additional 270,000 km^2 are being degraded by floods, salinity and alkalinity. An estimated 6,000 million tonnes of soil are lost every year from 800,000 km^2 alone. With the soil goes more than 6 million tonnes of nutrients which is more than the quantity applied in the form of fertilisers.

SIGNIFICANCE FOR OECD MEMBER COUNTRIES

In OECD Member countries which are net exporters of agricultural products and commodities some analysts argue that the agricultural land base is adequate to meet domestic food and fibre needs for many generations and, perhaps, indefinitely. However, this perspective does not adequately account for global interdependence in basic commodities and natural resources including, most notably, energy and food. If projected international food needs are taken into account, the future adequacy of the agricultural land resource base of these countries to satisfy all sources of demand without significantly higher real public and private costs of production, and without seriously degrading agriculture land, polluting ground water and so on, is - at best - uncertain. Those OECD Member countries which are net importers of agricultural products and commodities are even more vulnerable to the combined effects of increasing world demand for food and the increasing loss of arable land through conversion to non-agricultural uses in their own and in other countries.

Growing global interdependence for basic food, energy and other resource needs will increase the pressure on agricultural lands in all countries. Within OECD countries, the conversion of a small proportion of prime agricultural land to non-agricultural uses may lower current urban development costs and, for some, appear insignificant in contrast to the overall national endowment of agricultural resources. Yet the effects of such conversions are cumulative, and contribute to significant and avoidable reductions in agricultural potential and environmental quality.

For developing countries, any net losses and degradation of their arable land will, without compensating increases in agricultural productivity, increase their own demands and dependence on other countries (especially OECD Member countries) for additional food and increased financial and technical assistance. Having said this, it should be recognised that certain reductions in arable land are unavoidable and indeed necessary to development. Moreover, the potential for increasing crop yields is great in most developing countries.

The potential inflationary consequences of world food deficits in the international food market were noted in the report of the Brandt Commission, which further pointed out that "if additional assistance is not forthcoming, there will also be far more calls for emergency food supplies, which are in the long view an expensive and irrational way of coping with food problems. Food relief programmes often cost more in one year than would the five-year local investment programmes which might have made them unnecessary."[4]

The loss of cropland in developing countries is also due in part to demand originating in OECD Member countries as well as through the activities of some multinational enterprises headquartered in OECD Member countries.

CONSTRAINTS AND OPPORTUNITIES FOR FURTHER ACTION

The loss of arable land through conversion and soil degradation has not been perceived as a global issue of priority concern and this has been a constraint on international co-operation. Moreover, the consequences of future conversion are uncertain given the unpredictable nature of future trends in:

(i) world demand for agricultural commodities;

(ii) agricultural productivity;

(iii) climatic conditions;

(iv) resource quality and conservation;

(v) energy resource needs and opportunities;

(vi) domestic food and fibre consumption; and

(vii) urbanization.

---

(4) North-South: A Programme for Survival, London: Pan Books, 1980, p.94.

In developing countries there is a more fundamental and major constraint as the poor who comprise the majority of their populations often have little option but to sacrifice longer-term concerns and security for the sake of immediate economic and physical survival. The seriously poor in rural areas of developing countries constitute nearly a third of the world's population and, of those, 500 million suffer from malnutrition. In their efforts to satisfy their needs for food and fuel, the rural poor disrupt their own life-support systems, impair ecological processes and destroy genetic resources which limits, and sometimes forecloses, future possibilities for agricultural production. It is extraordinarily difficult to deal with these problems because of their huge scale, because there are so many people and production units with which to deal, and because of the pace of change. Because of the lack of economic flexibility among the rural poor, measures for the protection and conservation of agricultural land and soils will need to be complemented by measures that at least maintain and preferably improve their standards of living.

HIGHLIGHTS OF RELEVANT WORK AT INTERNATIONAL LEVEL

The world community is co-operating in considerable work relevant to this issue, largely through bodies of the United Nations system, and involving others such as IIASSA and ICSU.

The Annex reveals that the FAO is undertaking many relevant activities. For example, it is the lead agency in the development, adoption and follow-up of the World Soil Charter. In 1982, in co-operation with UNEP, UNESCO and the International Society of Soil Science (ISSS), the FAO will convene an international symposium to prepare a statement on world soil policy, and discuss the implementation of the plan of action for integrated soil resources management and protection.

The FAO (with UNEP, UNESCO, WMO, ISSS, IFIAS and ICSU) is preparing and testing an international system for the classification and monitoring of world soil resources. It is also (with UNEP, UNESCO, WHO, ISSS and IFIAS) conducting a survey of land transformation and its consequences for soil resources in vulnerable humid and arid tropical areas.

The 1979 Conference on Desertification resulted in a Plan of Action to Combat Desertification. Following the Conference, an Inter-Agency Working Group on Desertification was established, and efforts are now under way to establish it as an effective instrument of co-ordination for implementing the plan of action. In 1982-83, the FAO, UNEP and UNESCO will work together in an effort to improve monitoring methodologies for soil degradation, including desertification processes.

SOME THOUGHTS ON WHAT ELSE MIGHT BE DONE

In the context of increasing global interdependence, and uncertainty about its implications for agricultural land use in OECD Member countries, an early assessment of the extent, nature and rate of the conversion of agricultural land to non-agricultural uses in OECD Member countries would seem to be important. A complementary examination of the effects of selected national policies on the rate of conversion would also be useful. The central focus of this work should be the economic, environmental and social consequences of agricultural land conversion and soil degradation, together with the national and international implications of continued conversion of agricultural land to non-agricultural uses in OECD Member countries.

There is, of course, little by way of direct action that OECD itself can take concerning the loss of agricultural land and soil degradation in developing countries. It would be useful, however, to keep the situation constantly under review on behalf of the OECD Member countries which have the capacity for direct action through their bilateral assistance programmes and, indirectly, through their participation in other international organizations which have relevant programmes for dealing with these problems.

Part Three

MANAGEMENT ISSUES

Chapter VIII

ENVIRONMENTAL ASPECTS OF BILATERAL DEVELOPMENT CO-OPERATION

THE ISSUE

If bilateral development assistance policies are to contribute fully to sustainable economic development, they should seek to ensure that programmes and projects take greater account of their impact on the resources and the environment of recipient countries. At issue is how best to achieve this.

BACKGROUND

During 1978-1980, the International Institute for Environment and Development (IIED) conducted a detailed survey and analysis of the environmental policies and procedures of the bilateral aid agencies in Canada, the Federal Republic of Germany, the Netherlands, Sweden, the United Kingdom and the USA. The study was undertaken with the agreement, support and co-operation of these six bilateral aid agencies. The IIED and the six national research teams that carried out this study found that "there is general consensus in the aid agencies studied as to the meaning of 'environment' in the context of development problems. This represents a major change from the confused position of only three or four years ago. The most important feature of this consensus is that environment is now beginning to be seen not as an additional subject, the examination of which has to be added woodenly on to traditional development considerations; rather it is increasingly seen as a whole new approach to development which gives greater weight to the sustainability of results and to the costs of destructive side effects of projects." Another major finding of this study is that "this new view, however widely accepted theoretically, has still made too little impact on the orientation and design of the projects or practical development policies of the agencies studied."[1]

---

(1) Johnson, Brian and Blake, Robert, O., <u>The Environment and Bilateral Development Aid</u>, London: IIED, 1980, p.iii.

Two major concerns for development assistance are the link between poverty and environmental deterioration and the need to promote environmentally sound and sustainable economic development. The report of the Brandt Commission contains many references to these two concerns, including the following statement: "Concern for the future of the planet is inextricably connected with concern about poverty. Continued rapid population growth in the next century could make the world unmanageable; but that growth can only be forestalled if action is taken to combat poverty in this century. Much the same is true for the biological environment, which is threatened with destruction in many countries as a direct result of poverty - though in others as a result of ill-considered technological decisions and patterns of industrial growth. These problems can only be resolved by North and South acting in co-operation, and their mutual interests in doing so are only too obvious. The conquest of poverty and the promotion of sustainable growth are matters not just of the survival of the poor, but of everyone."(2)

In recognition of the importance of these concerns, ten major multilateral development institutions have already adopted a joint "Declaration of Environmental Policies and Procedures Relating to Economic Development". Among other commitments, they agreed to endeavour to institute procedures for the systematic examination of the environmental impact of all development activities under consideration for financing; to enter into co-operative negotiations with Governments and other agencies to ensure the integration of appropriate environmental measures in the design and implementation of economic development activities; and to provide technical assistance, including training, on environmental matters to developing countries.(3)

The OECD Environment Committee Meeting at Ministerial Level in May 1979 approved a "Declaration on Anticipatory Environmental Policies". The final operative paragraph declares that the Governments of OECD Member countries " will continue to co-operate to the greatest extent possible, both bilaterally and through appropriate international organisations, with all countries, in particular developing countries, in

---

(2) North-South: A Programme for Survival, London: Pan Books, 1980, p.75.

(3) Declaration adopted at the first meeting of the Committee of International Development Institutions on the Environment at UNDP, New York on 1st February, 1980.

order to assist in preventing environmental deterioration."(4)

The final section of the IIED report points out that the environmental performance of bilateral aid agencies could be improved simply through closer contacts and exchanges among them. It states: "If to date there has been surprisingly little contact among development aid agencies on questions of environmental protection and improvement, there has been only the most spotty and inadequate formal co-ordination. We believe that the question of improved inter-agency communication should be addressed by every development assistance organisation. The job is too big to be done alone or in an unco-ordinated fashion. One obvious forum already exists for such discussions, the Development Assistance Committee (DAC) of the OECD. Undoubtedly there could and should be discussions on environmental aspects of development aid under the auspices of this Committee."(5)

SIGNIFICANCE FOR OECD MEMBER COUNTRIES

Over 72 per cent of all official development assistance contributed in 1979 by OECD Members was in the form of bilateral aid. This represented more than half of the total net flow of concessional official assistance (approximately $16.4 billion out of some $30 billion in 1979) to developing countries and multilateral development institutions from all sources; i.e. the Development Assistance Committee (DAC), the Organisation of Petroleum Exporting Countries (OPEC), and the Council for Mutual Economic Assistance (CMEA). (6)

The impact of bilateral assistance provided by OECD Members is even larger than the above figures imply. The impact is larger in quantitative terms because of counterpart financing by the recipient countries. It is also larger because bilateral programmes influence the thinking, skills and capacities of key officials and managers in developing

---

(4) OECD and the Environment, 1979, page 24.

(5) Johnson, Brian and Blake, Robert, O., The Environment and Bilateral Development Aid, London: IIED, 1980, p.61.

(6) 1980 Review: Development Co-operation Efforts and Policies of the Members of the Development Assistance Committee, Paris: OECD, 1980.

countries whose decisions will affect their country's resources and environment for decades.

Impacts on OECD countries could take several forms. For example, if bilateral programmes or projects later turn out to have encouraged a form of development which was not sustainable, or which led to avoidable but by then serious or even irreversible environmental damage or deterioration, the donor countries could be faced with substantially increased requests for additional aid to deal with the newly created problems.

OECD countries may also soon find that their own economic development has been constrained (e.g. through the loss of genetic resources which might have had significant medical and commercial applications in their own countries -see Chapter VI). A similar point is made in the Report of the Brandt Commission, using deforestation as the example: "When the environment is overtaxed it does not harm only the countries directly faced with deterioration of the resource base but affects all countries through the ecosystem of the earth, as in the case of deforestation. The forests now covering about one-fifth of the earth's land surface are crucial to the stability of soil systems and to the survival of innumerable animal species and millions of human beings. They also help to absorb the excessive amounts of carbon dioxide emitted by burning fossil fuels, a process which threatens to warm up the atmosphere and could produce climatic change with potentially catastrophic consequences."[7]

CONSTRAINTS AND OPPORTUNITIES FOR FURTHER ACTION

A frequently cited and traditional constraint has been the delicacy of the relationship between the donor and recipient countries. Donor countries are generally neither willing nor able to impose specific environmental requirements if the recipient government is not convinced of the necessity for doing so. However, donor countries have at least some responsibility to ensure that significant environmental impacts and options for avoiding them have been adequately considered in the planning and implementation of major projects to which they contribute. Moreover, as pointed out by several of the experts consulted on this chapter, many developing countries are now seriously concerned about the environmental consequences of development projects and

---

(7) Op.Cit., p.113

would welcome assistance for developing a capacity to assess and avoid costly or irreversible environmental damage.

The need for a new approach for development assistance was also emphasized by Mr. Okita in the statement presented to the Environment Committee's April 1981 Special Session. He stressed that: "the content of aid needs to be re-assessed and greater importance attached to aid that helps environmental conservation and population control. Aid projects are usually chosen by recipient countries. A more positive approach seems necessary. Selection of the projects should be discussed by donor and recipient countries, taking into account the protection of the global environment. Population and resources problems are often connected with national sovereignty. This poses a delicate problem, but avoiding it may only worsen conditions in the future."[8]

A detailed review and analysis of the environmental procedures and practices of nine multilateral development agencies was published in 1979. It states that: "experience certainly shows that offers of help by aid agencies to ensure environmental soundness in projects are generally welcomed by recipient governments, if they avoid excessive interference with governments' development priorities and do not threaten to slow down their development process."[9]

Another major constraint is that project planners and managers in donor and recipient countries too often assume that environmental reviews or impact assessments will be too costly in time and money, and resist undertaking them fully or at all. However, the bilateral agency with the most extensive experience in this area, the US Agency for International Development (USAID), "estimates that the expense of preparing

---

(8) Okita, S., "Global Environmental Problems: Challenges and Responsibilities", in The Environment: Challenges for the '80s, OECD, Paris 1981 pp 21-22

(9) Stein, Robert, E., and Johnson, Brian, Banking on the Biosphere, Lexington, Massachusetts: D.C.Heath and Company, 1979, p.xiii.

environmental reviews amounts to a very small part of total project design and project costs."[10]

Many of the constraints are in the bilateral aid agencies themselves. As identified in the IIED study, these include the lack of adequately trained staff; of clear guidelines and criteria for evaluating projects before, during and after implementation; and of a clearly defined institutional focal point with responsibility for assuring the environmental soundness of projects.[11]

For those few bilateral aid agencies such as USAID which already undertake extensive enviromental reviews and impact assessments, there are many difficulties still to be overcome. The above cited review of USAID experience pointed out, for example, that:

(i) environmental assessments are generally too long and encyclopaedic rather than concise and analytical;

(ii) important secondary environmental impacts and longer term effects are usually not analysed adequately;

(iii) too many assessments are carried out with little host country participation; and

(iv) environmental analysis and monitoring is not extended often enough into the project implementation phase.

Some of the current needs and opportunities for further action include:

(i) the provision of special technical and other assistance to help developing countries evaluate their own resource and environmental problems, and to develop effective policies, strategies and institutional capacities for dealing with them more effectively;

---

(10) Blake, Robert O., et. al., Aiding the Environment: A Study of the Environmental Policies, Procedures and Performance of the US Agency for International Development, New York: Natural Resources Defense Council, 1980, pp.44.

(11) A detailed list of key functions to be performed by an environmental focal point is set out on page 59 of the IIED Report.

(ii) the development of general criteria to assist in designing specific types of environmentally sensitive projects (e.g. irrigation schemes) in order to eliminate or minimise environmental impacts; and

(iii) the convening of seminars and training courses for the staff of bilateral aid agencies on resource and environmental problems in developing countries, and on assessment methods and procedures, possibly including their counterparts from developing countries as well.

In considering all this a clear distinction needs to be drawn between the assessment of the environmental aspects of aid-assisted development projects and projects specifically designed and undertaken to serve environmental protection or resource conservation objectives (e.g., reforestation or wildlife management projects). The latter could raise a range of different policy issues not considered here.

HIGHLIGHTS OF RELEVANT WORK AT INTERNATIONAL LEVEL

As noted above, in February 1980 ten major multilateral agencies(12) adopted a joint "Declaration of Environmental Policies and Procedures Relating to Economic Development". Since then, representatives of the signatories have met twice to consider ways of extending application of the principles. A third meeting was held in Brussels in April, 1982, to which bilateral agencies were invited.

In November 1981, the High-Level Meeting of the OECD Development Assistance Committee (DAC) agreed to convene a special meeting on Aid and Environmental Protection. The meeting was intended to enable an exchange of views on the policies and procedures established by aid agencies to deal with environmental concerns, and the lesson to be learnt from their practical implementation. The meeting, held in April

---

(12) The African Development Bank; The Arab Bank for Economic Development in Africa; the Asian Development Bank; the Caribbean Development Bank; the Inter-American Development Bank; the World Bank; the Commission of the European Communities; the Organisation of American States; the United Nations Development Programme; and the United Nations Environment Programme.

1982, undertook to provide a factual survey of these policies and procedures in Member countries and to identify issues with implications for project selection, design and appraisal.

In April 1981, the International Union for the Conservation of Nature and Natural Resources (IUCN) established the Conservation for Development Centre (CDC) with the aim of providing support and technical advice on the incorporation of conservation considerations in the development programmes and projects of governments and development agencies. It will also maintain a roster of experts available to work or advise on such projects.

The role of consulting organisations in planning, designing and implementing international development assistance projects that affect the environment, and in carrying out environmental assessments, has been receiving increased attention. In 1981, the IIED published a report commissioned by UNEP on "Environmental Performance of Consulting Organisations in Development Aid." In March 1981, a major conference on the environmental practices of consulting firms working on international development projects was convened in New York by the World Environment Centre, in co-operation with the World Bank and the US Agency for International Development.

SOME THOUGHTS ON WHAT ELSE MIGHT BE DONE

In light of the above, it is evident that a great deal will need to be done, over time, to further the integration of environmental considerations in bilateral development assistance policies, programmes and projects. As an important step in this direction, bilateral development assistance agencies, working through DAC and other appropriate bodies of OECD, could develop a set of guidelines, and a procedure for periodically reviewing and revising them, under which they would, for example, undertake:

(i) to ensure that significant environmental effects and options for avoiding them have been adequately considered in the design and implementation of projects to which they contribute; and to ensure in particular that such projects do not contribute to further unnecessary loss or degradation of soils, genetic resources and tropical forests. (See Chapters VI and VII);

(ii) to review periodically the application of the guidelines in the policies, programmes and projects of bilateral assistance agencies;

(iii) to provide additional technical and other assistance to help developing countries to develop effective policies, strategies and institutional capacities for dealing with their environment and resource management problems; and

(iv) to convene special seminars and training courses (on, for example, major resource and environmental problems in developing countries and project appraisal and environmental assessment methods and procedures) for the staff of bilateral assistance agencies and preferably their counterparts in developing countries as well.

As a part of the process of developing such guidelines, it would be useful to convene a special panel or seminar on the implications for bilateral assistance programmes of the environment and resource issues raised in this volume and in the major recent reports on which they are based.

The practice of appraising bilateral aid projects and programmes is assisted by the availability of a number of well-known reference documents (such as the "Manual of Industrial Project Analysis in Developing Countries" published by the OECD Development Centre in 1969[13]). In view of the time and the context of this publication, and perhaps others, the external effects of projects, especially resource and environmental effects, were only considered to a limited extent. Since then, these elements have become far more significant considerations in project appraisal. Revised versions of these references or perhaps a similar but new manual, could usefully be prepared and issued.

---

(13) Little, I.M.D., and Mirrlees, J.A. <u>Manual on Industrial Project Analysis in Developing Countries, Volume II: Social Cost Benefit Analysis,</u> Paris: OECD, 1969.

Chapter IX

ENVIRONMENTAL IMPACT ASSESSMENT AND INTERNATIONAL CO-OPERATION[1]

THE ISSUE

Environmental aspects need to receive more adequate and earlier consideration in the design, planning and implementation of major policies, programmes and projects, especially those which are international or have international implications. This will require improvements in the methodology, procedures and use of environmental impact assessments

BACKGROUND

The Environment Committee Meeting at Ministerial level in 1974, adopted a Recommendation, subsequently endorsed by the OECD Council, on "The Analysis of the Environmental Consequences of Significant Public and Private Projects."[2] At the Committee's second Meeting at Ministerial level in May 1979, it was noted that "since 1974, major efforts had been made in most Member countries to make more extensive use of environmental impact assessments, to improve the methodologies, to make more widely available to the public studies on specific projects and to introduce or re-inforce legal requirements to prepare impact assessments either by specific laws or within existing regional, urban and utility planning legislation and decision structures. Although these achievements over a period of years were considered remarkable, it was generally recognised that there were too often inadequacies in procedures, gaps in methodologies and shortcomings in the assessment techniques which may require further work and co-operation at the international level."

---

(1) The use of environmental impact assessments in bilateral assistance programmes and projects is discussed briefly in Chapter VIII.

(2) C(74)216

Consequently, at this same meeting, Ministers adopted a Recommendation on "The Assessment of Projects with a Significant Impact on the Environment". In endorsing it, the OECD Council[3], at the same time, instructed the Environment Committee "to report on the practical experience of Member countries in implementing various assessment methods and procedures and on actions taken pursuant to this Recommendation."

The Recommendation provided that Member governments consider environmental assessment procedures for actions that might have significant transboundary effects. There are international applications of environmental impact assessments, however, that are not covered by that Recommendation. While the Recommendation may possibly cover the impact of a project in one OECD country on another OECD country, it clearly does not cover assessment of projects or activities undertaken by a Member country that may reasonably be expected to have an adverse impact on the environment or ecology of a non-OECD country hosting the project, on the environment of countries neighbouring it, or on the global commons.

The recent report of the Brandt Commission dealt with this question and proposed that: "Environmental impact assessment should be undertaken wherever investments or other development activities may have adverse environmental consequences whether within the national territory concerned, for the environment of neighbouring countries or for the global commons. There should be guidelines for such assessments, and when the impact falls on other countries there should be an obligation to consult with them".[4]

OECD Member countries have agreed to certain guidelines for the conduct and exchange of environmental impact assessments for actions with international implications, but these apply only to problems of transfrontier pollution among OECD countries. The guidelines include, for example, the provision that "the country of origin should communicate to the exposed country the results and elements that are pertinent for questions of transfrontier pollution of environmental impact studies..."[5]

---

(3) C(79)116

(4) North-South: A Programme for Survival, London: Pan Books, 1980, p.115.

(5) C(78)77 (Final), Annex, Part II, para.3.

SIGNIFICANCE FOR OECD MEMBER COUNTRIES

OECD Member countries took the lead in the early and mid-1970s in initiating environmental impact assessments. Since then they have acquired a great deal of practical experience in applying and improving them. Further improvements in the methodology and procedures not only will benefit OECD Members but also will be of interest and use to many countries outside the OECD region as well as bilateral and multilateral development assistance institutions.

Certain economic, trade, energy and other policies of OECD countries may have significant environmental impacts but they remain largely unassessed. The international environmental impact of national environmental policies themselves is also largely unassessed or neglected. In his April 1981 statement to the Special Session of the OECD Environment Committee, Mr. Okita said: "an action that may be reasonable for protecting one country's environment may not be for protection on a global scale (e.g., felling of tropical forests overseas because domestic forests are protected). The sum of national environmental protection policies cannot therefore be sufficient to cope with global environmental problems."(6)

CONSTRAINTS AND OPPORTUNITIES FOR FURTHER ACTION

A major and continuing constraint which applies generally to environmental impact assessment is "the lack of rigorous and unambiguous methods to (a) identify the risks, costs and benefits involved in a particular policy option; (b) measure the identified risks, costs and benefits; and (c) attach specific values to the risks, costs and benefits so that they can be compared to determine whether the policy option should be adopted on the basis that the benefits outweigh the risks and costs."(7)

---

(6) Okita, S., "Global Environmental Problems: Challenges and Responsibilities", in _The Environment: Challenges for the '80s_, Paris: OECD, 1981, p.19.

(7) _Survey of Science and Technology Issues Present and Future_, Staff Report of the Committee on Science and Technology, U.S. House of Representatives, Washington: U.S. Government Printing Office, 1981, p.180.

Another constraint stems from the fact that very often projects are assessed only after site selection, planning and design have been completed. It is an "add-on" process and procedure which, by definition, involves delays in decision and implementation whether or not changes in site and plans are required. These delays can be costly. The alternative of building environmental assessment into the siting, planning and design of projects is difficult. Experience demonstrates, however, that it is usually far more effective environmentally and economically to do so.

The OECD Council has, in fact, recommended that Member governments "co-ordinate the procedures for, and the form and timing of, the appropriate assessment of the environmental impact as an input to the planning and decision-making process, thus ensuring early consideration of measures for mitigating environmental impact, for enhancing environmental quality, and for avoiding undue delays of projects."(8)

HIGHLIGHTS OF RELEVANT WORK AT INTERNATIONAL LEVEL

As the Annex indicates, the United Nations Environment Programme has several relevant activities under way or planned. These include the preparation of global guidelines on environmental impact assessment; the development of environmental criteria for the siting of industry; and a series of case studies of natural disasters and ecological mismanagement. The latter is being undertaken in co-operation with the IUCN and the Red Cross.

UNEP and other members of the Committee of International Development Financing Institutions on Environment also review at their annual meetings the incorporation of environmental considerations in the projects they support.

The United Nations Economic Commission for Europe (ECE) is examining the experience of ECE Governments in anticipating and assessing the environmental consequences of economic activities and technological development, especially in connection with transboundary problems.

Of all international agencies, the World Bank has perhaps the most highly developed methods and procedures for assessing and avoiding significant environmental impacts. It has issued relevant guidelines and a number of reports on environmental assessment.

---

(8) C(79)116, Part I, operative para.2.

SOME THOUGHTS ON WHAT ELSE MIGHT BE DONE

As noted earlier, there is a growing concern about delays in decision-making, and about streamlining procedures for environmental assessments. This presents both a need and opportunity for further international co-operation in this field.

In this connection, it would be useful to explore the legal, institutional, economic and other means available to encourage, induce and enable public and private bodies to make environmental assessment an integral part of their decision-making.

It would also be useful to develop guidelines concerning environmental assessments for projects that may reasonably be expected to have adverse consequences on the environment or ecology of a country hosting the project, on the environment of countries neighbouring it, or on the global commons.

Chapter X

ENVIRONMENTAL ASPECTS OF MULTINATIONAL INVESTMENT

THE ISSUE

The issue for OECD countries is how best to develop and encourage the application of guidelines for multinational enterprises designed to ensure that such firms, in performing their activities, take appropriate steps to protect the environment.

BACKGROUND

Foreign investment, including investment through multinational enterprises, plays a major role in global economic development and trade, both in OECD economies and in those of the Third World.

During the past decade, 60 per cent of the industrial investment in Third World countries originated outside those countries, and much of this was made by multinational enterprises. A large proportion went into the exploitation of natural resources (e.g. minerals, fuels, timber, fish) for use largely by OECD economies. This will continue. The World Bank estimates world-wide capital requirements for minerals between 1977 and 2000 at $278 billion in constant dollars. Of this, $96 billion would be invested in the Third World and, of this, three-quarters would need to be financed from external sources. In addition, industrialized countries are expected to make heavy investments to develop deep sea bed mining.

Multinationals are also responsible for a major part of the growing volume of trade in minerals, manufactures and technology between OECD and the Third World. In 1978, this trade represented 25 per cent of total trade by OECD countries.

While this investment and trade plays an essential and largely beneficial role in development, there is no doubt that much of it has been associated with heavy inroads on exhaustible resources and avoidable damage to the environment and to the essential ecological basis for sustainable economic growth. Examples abound not only in forestry, mineral and water developments but also in the deployment of inappropriate technologies and in the marketing of products harmful to man and the environment. Moreover, there is evidence of a trend to site some primary industries in developing countries to take advantage of lower labour costs and other production costs. While there are excellent reasons why this trend should not be discouraged, it is of interest to note that it seems to be particularly strong in respect of certain traditional and heavily polluting industries such as steel, aluminium, asbestos and certain toxic chemicals.(1)

Rightly or wrongly, OECD countries may feel relatively confident in their ability to ensure that investment, technology and trade, whatever their source, will not do great harm to their environment. This may not be the case, however, for many regions in the Third World. Environmentally unsound development confronts many of these countries with difficult trade-offs between the immediate benefits of development and its long-term costs to human health and the environment. Coupled with the lack of effective means to manage resources and the environment and to assess the implications of development proposals, the result can be extensive and often irreparable damage to the ecological basis both for the development concerned and for future economic development.

SIGNIFICANCE FOR OECD MEMBER COUNTRIES

The consequences of ecological deterioration stemming from environmentally unsound investments in developing countries are felt not only in the recipient country, but also in neighbouring countries and, in fact, can also impact back on OECD countries. The latter is discussed vividly in the Okita report.(2) The Brandt Commission has pointed out that 60 per cent of the world exports of major agricultural and mineral commodities other than oil originate from the Third World. Both the EEC and the USA depend entirely on the Third World for imports of

---

(1) Castleman, Barry, The Export of Hazardous Factories to Developing Nations, Washington: March 1978.

(2) Basic Directions in Coping with Global Environmental Problems; Ad hoc Group on Global Environmental Problems; Japan: Dec. 20, 1980.

tropical hardwoods, rubber, jute, bananas, tea, cocoa and coffee. Japan and the EEC obtain 90 per cent of their supplies of many important minerals from the Third World.(3)

Thus, while depletion of tropical forests, loss of soil and soil fertility, loss of unique genetic materials and persistence of harmful pesticides residues in food hit hardest the people and economies of the countries concerned, the interdependent economies and peoples of OECD do not escape.(4) Moreover, ecological deterioration and resource exhaustion stemming from environmentally unsound development may create a need for greatly increased development assistance to overcome problems that could have been avoided. As the Brandt Commission stated: "to seek to attract industry at the expense of the environment might cause damage that is more costly to undo than to prevent".(5)

HIGHLIGHTS OF RELEVANT WORK AT INTERNATIONAL LEVEL

The OECD adopted a set of "Guidelines for Multinational Enterprises" in 1976 and revised them in 1979. These guidelines, however, do not include a chapter on environmental protection. Consequently, consideration is now being given in OECD to developing appropriate guidelines for environmental protection which could be taken into account when the present guidelines are reviewed again in 1984 by the OECD Council.

An Inter-governmental Working Group of the ECOSOC Commission for Transnational Corporations is now developing a comprehensive draft code of conduct for transnational corporations. The draft includes a section on environmental protection which includes clauses on, for example: measures to avoid and remedy environmental damage; the application of adequate technologies; and the provision to importing countries of relevant information on the characteristics of products, processes and other activities which may harm the environment; on the measures and costs involved in avoiding or mitigating their harmful effects; and on

---

(3) North-South: Programme for Survival, London: Pan Books, 1980, p.72.

(4) See Chapters VI and VII.

(5) Ibid., p.115.C.

any prohibitions, restrictions imposed vis-a-vis these products, processes and services in other countries. It is expected that a completed draft will be submitted for consideration by the ECOSOC Commission for Transnational Corporations in the latter half of 1982.

Chapter XI

INTERNATIONAL APPLICATION OF THE POLLUTER-PAYS PRINCIPLE

THE ISSUE

The issue is whether and how the Polluter-Pays Principle can and should be extended to cover one or both of (a) the case of investment within an OECD country that gives rise to spillovers into the global commons, and (b) the case of investment abroad that gives rise to spillovers within the non-OECD recipient country, into neighbouring countries, or into the global commons.

BACKGROUND

In 1972 the Polluter-Pays Principle was adopted by the Council of the OECD as part of the Guiding Principles Concerning International Economic Aspects of Environmental Policies.[1] The principle reflected a conviction that economic efficiency would be promoted and distortions of trade avoided if the pollution control policies of Member countries required polluters to internalise their external costs. In effect, while recognising certain exceptions during a transition period, the principle was intended to ensure that the full cost of pollution control or prevention measures should be reflected, in turn, in the costs of the goods produced and marketed.

The principle was adopted by OECD Member countries in order to provide a common and effective economic guideline on which to base their pollution control policies. The Polluter-Pays Principle has been the subject of gradual implementation by OECD Member countries. There continue to be exceptions or special arrangements in implementing the PPP in some countries, but these are permitted only as long as they do not lead to significant distortions in international trade and investment.

---

(1) C(72)128

The application of the Polluter-Pays Principle is presently required for investment in, and operation and maintenance of, pollution abatement facilities and measures undertaken by the public and private sectors within each OECD Member country.

OECD's Polluter-Pays Principle does not at present apply to at least two cases which are at issue here.

- First, it does not apply in the case of investment within an OECD country that gives rise to spillovers into the global commons, unless an international agreement to reduce pollution from the source exists;

- Second, it does not apply in the case of investment abroad that gives rise to spillovers from the country hosting the investment into countries neighbouring it or into the global commons, where the country hosting the investment does not apply the PPP.

However, the OECD Council recently adopted a Recommendation "Concerning Certain Financial Aspects of Actions by Public Authorities to Prevent and Control Oil Spills". According to this Recommendation, governments should apply the Polluter-Pays Principle and assign to the person or entity liable for the spill the costs of reasonable remedial action taken by public authorities after an incident.(2)

SIGNIFICANCE FOR OECD MEMBER COUNTRIES

Significant pollution of the global commons originates from activities of OECD Member countries. This includes, for example, the dumping of radioactive and toxic waste into the ocean; the release of persistent chemicals in enclosed or semi-enclosed seas; emissions of chemicals which may be detrimental to the ozone layer; and emissions which may affect the world climate. Efforts to address these issues to date have been ad hoc and focussed largely on legal remedies. There would seem to be considerable merit in attempting to address these issues also from the perspective of economic efficiency at the global level.

---

(2) C(81)32 Final. The question of applying the PPP in the case of transfrontier pollution was discussed by the OECD in 1979. See reports on international financial transfers published in "Transfrontier Pollution and the Role of States", OECD, 1981.

Most of the international investment in the world also originates in OECD Member countries. These investments, while contributing to the creation of goods, services and jobs in other countries, may also impose environmental costs on those countries and their neighbours as well as on the global commons. The environmental constraints applying to such investments, the environmental degradation which they can cause (even when this is in line with internal environmental law) and the subsidies which may be offered to cover anti-pollution costs, would need to be examined from the point of view of the host country, the country from which the investment originates, and the other countries possibly affected.

The current climate with its emphasis on efficiency and cost-effectiveness would appear to offer a special opportunity to address this growing problem. The dangers of not doing so were emphasised by Mr. Okita in the statement presented to the special OECD meeting in April 1981 on "OECD and Policies for the '80s to Address Long-term Environment Issues". He pointed out that "although it is necessary that each country implements its domestic environmental policies, problems of a global scale cannot be solved by the mere accumulation of domestic environmental policies by individual countries. For any country to discharge pollutants into the environment - common property of the whole planet - is an 'extranational' diseconomy, just as the external diseconomy of a business in discharging pollutants is not reflected in its costs. When this happens, that country's activities tend to exceed the level that is most appropriate when external costs are internalized, and are apt to destroy the environment on a global scale."(3)

SOME THOUGHTS ON WHAT ELSE MIGHT BE DONE

There does not appear to be any international work under way or planned relevant to the international application of the Polluter-Pays-Principle. If any further work were to be undertaken, however, it could examine the feasibility of extending the scope of the Polluter-Pays-Principle to at least the following two cases:

(i) the case of investment within each country which gives rise to significant environmental impacts on the global commons;

---

(3) Okita, S., See "The Environment: Challenges for the '80s", OECD 1981.

(ii) the case of investment abroad that gives rise to significant environmental impacts within the host country, in neighbouring countries, or on the global commons.

Annex I

HIGHLIGHTS OF RELEVANT ACTIVITIES UNDER WAY OR PLANNED BY INTERNATIONAL ORGANISATIONS ON SELECTED ENVIRONMENT AND RESOURCE ISSUES

This annex is intended to provide a summary overview of the highlights of the activities of key international organisations relevant to the issues treated in this report. It is necessarily selective, and does not include references to all possibly relevant projects by every major organisation.

A draft of this Annex was circulated in October 1981 to the international organisations mentioned for their comments and now reflects suggestions received.

The format has been organised so that the issues are presented in the same sequence as in the report.

Each entry includes a short description of the relevant programme or project; an indication of its starting date or duration (a "C" indicates a continuous activity); and the abbreviated title of the sponsoring organisation (s). The abbreviated title and corresponding full title for each organisation cited in the report and Annexes is given in Annex II. Annex III is a list of relevant global monitoring and information systems.

TABLE OF CONTENTS

HIGHLIGHTS OF RELEVANT ACTIVITIES OF INTERNATIONAL ORGANISATIONS

ENVIRONMENTAL POLLUTION ISSUES

CO2 and Climatic Changes

Activity	Organisations
- Strengthen the WMO global system for climatological observation, data collection and retrieval from national stations as part of WCIP, the World Climate Impact Studies Programme (C)	WMO UNEP
- Conduct assessment of impacts due to CO_2-induced climatic changes on agriculture	WMO,UNEP SCOPE,WHO,
- Fisheries, water supply, sea level, energy, health, and natural ecosystems (C)	UNESCO,IOC FAO, IEA
- Study effects of improved photosynthesis efficiency on carbon cycle, and atmospheric CO_2 (83)	UNEP,WMO UNESCO, IUCN,SCOPE
- Study the carbon cycle, especially issue of CO_2 and climate (82-83)	UNEP,SCOPE UNESCO IOC
- Conduct national assessments on responses of economic/social/political system to climatic changes (82-83)	UNEP ICSU
- Convene a scientific conference to develop an Action Plan for assessment of CO_2 - induced climatic change (82)	WMO,UNEP FAO,ICSU

The Ozone Layer

Activity	Organisations
- Carry out the following projects under the auspices of CCOL-Co-ordinating Committee on the Ozone Layer:	
- Determine biological effects of UV-B on terrestrial and aquatic ecosystems and agriculture (82)	ICSU UNEP
- Provide continuing measurements of integrated ozone layer and depletion (C)	WMO
- Correlate UV-B measurements with human health effects (C)	WHO WMO

- Conduct preparatory work for a global convention for the protection of the ozone layer (C) — UNEP, WMO, WHO, CEC

Acid Precipitation

- Under the EMEP, the Co-operative Programme for monitoring and evaluating the long-range transmission of air pollutants in Europe, conduct an overall assessment of pathways, deposition, concentration and effects of sulphur compounds on human health, living resources, ecosystems, soils, vegetation, materials and visibility, leading to a review under the terms of the Convention on Long-Range Trans-boundary Air Pollution (81-83) — UNEP, ECE, WMO

- Study the sulphur cycle, including problems of urban area pollution and eutrophication of water bodies (81-83) — UNEP, ICSU

- Study use of chemical treatments for atmospheric pollution by sulphur oxides (82-83) — ECE

- Convene Third Seminar on Desulphurization of Fuels and Combustion Gases (1981) — ECE

- Review SO_x strategies and policies and prepare guidelines on sulphur emission reduction strategies, with special reference to developing countries (82-83) — UNEP, REC's

- Study impact of acid rain on freshwater and terrestrial ecosystems (83-85) — UNESCO

Chemicals

- Prepare technical reports and guidelines on general principles and methods on testing and evaluating food additives and contaminants; environmental and health monitoring; early detection of health impairment in occupational health; chemical and biochemical methods for assessing the hazard of pesticides to man; and establishment of permissable levels of occupational exposure (C) — WHO, FAO

- Under the "Environmental Health Criteria Programme", assess relationship between exposure to environmental agents and human health; provide guidelines for setting — WHO

Activity	Agency
exposure limits; and preliminary reviews of chemicals and other agents likely to be used increasingly in industry, agriculture, or at home (C)	
- Prepare and distribute data profiles for chemicals listed in the IRPTC (82-83)	UNEP
- Convene symposium to prepare recommendations on methodologies to measure and evaluate effects of chemicals on man standards (82)	UNEP
- Determine ecological effects of pesticides and fertilisers (C)	UNESCO FAO IUCN
- Implement (IPCS) the International Programme on Chemical Safety on risk assessment of chemicals and evaluation of specific toxicological effects and elaboration or updating of methodology (C)	UNEP WHO ILO
Hazardous Wastes	
- Convene workshop on policy guidelines and a code of practice for handling hazardous wastes (81-82)	WHO UNEP
- Collect and analyse information on the economic, technical, environmental and social impact of existing international exchanges of waste and investigate possibilities for expanding waste exchange networks (C)	ECE
- Prepare guidelines on trans-boundary transport and disposal of chemical wastes (82)	UNEP WHO
- Consider policy measures for handling, control and disposal of toxic wastes, including the reduction of toxic wastes through the application of low and non-waste technologies and through re-utilization and re-cycling (C)	ECE
- Prepare a manual on environmental management in the chemical industries (82)	UNEP
- Study management of wastes from uranium mining and milling and from nuclear power plants (81-83)	IAEA

	Action	Organisations
-	Review toxicological problems associated with specific industries and technologies (82-83)	WHO UNDP

RESOURCE ISSUES

Maintaining Biological Diversity

	Action	Organisations
-	Implement the World Conservation Strategy to maintain essential ecological processes and life-support systems; to preserve genetic diversity; and to ensure the sustainable utilization of species and ecosystems (C)	IUCN UNEP WWF FAO UNESCO
-	Encourage governments to ratify international conservation treaties (including the Convention on International Trade in Endangered Species of Wild Fauna and Flora) and assist them in formulating appropriate national legislation and regulations (C)	IUCN UNEP WWF FAO
-	Conduct systematic monitoring, issue assessment statements on status of plant and animal species, and prepare action plans for the conservation of wild genetic resources (C)	IUCN UNEP FAO UNESCO
-	Assist governments in the preparation of national conservation plans and strategies(C)	IUCN UNEP FAO
-	Conduct global programme for establishing Biosphere Reserves (209 to date) for the conservation of natural areas and their genetic material (C)	UNESCO
-	Evaluate international measures taken or being considered for the conservation of flora, fauna, and their habitats in the ECE region; and study the possible promotion of a network of representative ecological areas in the ECE region (81-84)	ECE CE UNESCO
-	Collect specified germplasms for crop plants according to approved priority list; evaluate and store them; and publish information on them and on methodologies for on-site conservation (C)	FAO,UNEP UNDP IBPGR CGIAR UNESCO
-	Continue survey of distribution of fish populations in biosphere reserves and in specific marine and inland water systems,	UNESCO FAO UNEP

Activity	Agency
and develop methodologies for assessing genetic impoverishment in selected threatened fish populations (82-83)	IUCN
- Continue world assessment of forest resources and examine in detail the nature, rates and causes of change in tropical forests (C)	FAO UNEP UNESCO
- Study trends regarding the removal of tropical forest cover and its replacement through various forms of afforestation and reforestation (C)	FAO UNEP UNESCO
- Conduct case studies and develop further guidelines and management tools for regular and sustained production in indigenous tropical forests and woodland ecosystems (82-83)	FAO UNESCO UNEP IUCN
- Develop principles, criteria and guidelines for selection, establishment and management of tropical forest areas, and identify research priorities (83)	IUCN UNESCO FAO UNEP
- Promote integrated conservation programmes for tropical moist forests in selected selected countries in Africa, Asia and Latin America (82-83).	IUCN WWF
- Assess the ecological effects of increasing human activities on tropical and sub-tropical forest ecosystems (C)	UNESCO
- Promote and assist in establishing legal regimes for sound tropical forest management and protection (82-83)	FAO REC's UNEP,IUCN
- Examine application of techniques of cost/benefit evaluation for tropical deforestation in Latin America (82-83)	UNEP ECLA
- Investigate methodologies for surveying and indexing genetic impoverisment within woody species; publish the data; and promote on-site conservation (82-83)	UNEP, UNESCO, FAO, IUCN IBPGR
- Assess germplasm conservation of multi-purpose tree species (82-83)	FAO,UNEP ICRAF
- Promote research on genetic impoverishment and long-term storage of tree germplasm (C)	FAO,UNESCO UNEP,IUFRO

Loss of Cropland and Soil Degradation

Action	Agencies
- Convene group of experts to prepare statement on world soils policy and implement the Plan of Action for integrated soil resources management and protection (82)	FAO UNEP UNESCO ISSS
- Prepare and test an international system and criteria for the classification and monitoring of world soil resources (C)	FAO,UNEP UNESCO,WMO ISSS,IFIA ICSU
- Undertake case studies of soil conservation problems in selected regions, including identification of social, economic and legislative constraints (C)	FAO,UNEP UNESCO IFIAS,ICSU IIASA, CGIAR
- Conduct surveillance of land transformation and its consequences for soil resources in vulnerable humid and arid tropical areas (C)	FAO, UNEP UNESCO,WHO ISSS,IFIAS ICSU
- Improve monitoring methodologies for soil degradation, including desertification processes (82-83)	FAO UNEP UNESCO
Develop and apply methodology for assessment and mapping of desertification hazards, and test feasibility of world desertification hazards map (82-83)	FAO UNESCO WMO ISSS
- Convene an international symposium to exchange information on regional development plans and strategies for arid zones in developing countries (82)	UNEP UNIDO FAO UNESCO
- Establish the IAWGD, the Inter-Agency Working Group on Desertification, as an effective and main instrument of co-ordination for implementing the 1977 Plan of Action to Combat Desertification (C)	UNs
- Secure adoption and follow-up for the World Soil Charter (C)	FAO

MANAGEMENT ISSUES

Environmental Aspects of Bilateral Development Co-operation

Action	Agencies
- Extend application of Declaration of	UNEP

Principles on incorporating environmental considerations in projects to other multilateral and bilateral agencies (C)	IIED
- Provide support and technical assistance for the incorporation of conservation considerations in the development programmes and projects of governments and development agencies through the Conservation for Development Centre (C)	IUCN

Environmental Impact Assessment and International Co-operation

- Review, at annual meetings of Committee of International Development Financing	UNEP,UNDP World Bank
Institutions on Environment, the incorporation of environmental considerations in funded projects (C)	AfDB,AsDB IADB, CDB ABEDA,EDF, OAS
- Prepare global guidelines on environmental impact assessment (82-83)	UNEP
- Study procedures and relevant experience of ECE Governments in anticipating and assessing the environmental consequences of economic activities and technological development, especially in connection with trans-boundary problems (82)	ECE
- Undertake case studies of natural disasters and ecological mismanagement (C)	IUCN UNEP Red Cross
- Study environment impact of inter-basin water transfers and other major water projects (C)	UNESCO
- Development and application of guidelines on industrial environmental impact assessment and environmental criteria for the siting of industry (82-83)	UNEP
- Elaborate a cost effective and practical system for environmental impact assessment to assist developing countries (82-83)	UNEP

Environmental Aspects of Multinational Investment

- The Intergovernmental Working Group on a Code of Conduct of the ECOSOC Commission on Transnational Corporations is scheduled	UNCTC

to submit a draft code, including a section on environment, for consideration by the Commission in 1982.

Annex II

ABBREVIATIONS USED IN REPORT AND ANNEX I

ABEDA	Arab Bank for Economic Development in Africa
AfDB	African Development Bank
AsDB	Asian Development Bank
CCOL	Co-ordinating Committee on the Ozone Layer
CDB	Caribbean Development Bank
CE	Council of Europe
CEC	Commission of the European Communities
CGIAR	Consultative Group on International Agricultural Research
CNRET	United Nations Centre for Natural Resources, Energy and Transportation
DAC	Development Assistance Committee (OECD)
ECA	Economic Commission for Africa
ECE	Economic Commission for Europe
ECLA	Economic Commission for Latin America
ECOSOC	United Nations Economic and Social Council
EDF	European Development Fund
ESCAP	Economic Commission for Asia and the Pacific
FAO	Food and Agriculture Organisation of the United Nations
IADB	Inter-American Development Bank
IAEA	International Atomic Energy Agency
IAWGD	Inter-Agency Working Group on Desertification

IBPGR	International Board for Plant Genetic Resources
ICRAF	International Council for Research in Agro-Forestry
ICSU	International Council of Scientific Unions
IEA	International Energy Agency
IFIAS	International Federation of Institutes for Advanced Studies
IIASA	International Institute for Applied Systems Analysis
IIED	International Institute for Environment and Development
ILO	International Labour Office
IOC	Inter-Governmental Oceanographic Commission
IPCS	International Programme on Chemical Safety
ISSS	International Society of Soil Science
IUCN	International Union for the Conservation of Nature and Natural Resources
IUFRO	International Union of Forestry Research Organisations
OAS	Organisation of American States
OECD	Organisation for Economic Co-operation and Development
RECS	United Nations Regional Economic Commissions
S/CITES	Secretariat for the Convention on International Trade in Endangered Species of Wild Fauna and Flora
SCOPE	Scientific Committee on Problems of the Environment
UNCTC	United Nations Centre on Transnational Corporations
UNDP	United Nations Development Programme
UNEP	United Nations Environment Programme

UNESCO	United Nations Education, Scientific and Cultural Organisation
UNIDO	United Nations Industrial Development Organisation
UNS	United Nations System
WCIP	World Climate Impact Studies Programme
WHO	World Health Organisation
WMO	World Meteorological Organisation
WWF	World Wildlife Fund

Annex III

LIST OF GLOBAL MONITORING AND INFORMATION SYSTEMS

Acronym	Title	Agency
AGRIS	International Information System for the Agricultural Sciences and Technology	FAO
ASFIS	Aquatic Sciences and Fisheries Information System	FAO
CMC	Conservation Monitoring Centre	IUCN
ELIS	Environmental Law Information System	IUCN
GEMS	Global Environmental Monitoring System	UNEP
IGOSS	Integrated Global Ocean Station System	IOC
INFOTERRA	International Referral System for Sources of Environmental Information	UNEP
INRES	Information Referral System for Technical Co-operation among Developing Countries	UNIDO
	International Occupational Safety and Health Hazard Alert System	ILO
IRPTC	International Register of Potentially Toxic Chemicals	UNEP
MARPOLMON	Marine Pollution Monitoring Programme	IOC
MEDI	Marine Environmental Data and Information System	IOC
UNISIST	Inter-governmental Programme for Co-operation in the field of Scientific and Technological Information	UNESCO

WDC	World Data Centre on Micro-organisms	UNESCO
WWW	World Weather Watch	WMO

MARIA GREY
LIBRARY
W.L.I.H.E.

OECD SALES AGENTS
DÉPOSITAIRES DES PUBLICATIONS DE L'OCDE

ARGENTINA – ARGENTINE
Carlos Hirsch S.R.L., Florida 165, 4° Piso (Galería Guemes)
1333 BUENOS AIRES, Tel. 33.1787.2391 y 30.7122

AUSTRALIA – AUSTRALIE
Australia and New Zealand Book Company Pty, Ltd.,
10 Aquatic Drive, Frenchs Forest, N.S.W. 2086
P.O. Box 459, BROOKVALE, N.S.W. 2100

AUSTRIA – AUTRICHE
OECD Publications and Information Center
4 Simrockstrasse 5300 BONN. Tel. (0228) 21.60.45
Local Agent/Agent local :
Gerold and Co., Graben 31, WIEN 1. Tel. 52.22.35

BELGIUM – BELGIQUE
LCLS
35, avenue de Stalingrad, 1000 BRUXELLES. Tel. 02.512.89.74

BRAZIL – BRÉSIL
Mestre Jou S.A., Rua Guaipa 518,
Caixa Postal 24090, 05089 SAO PAULO 10. Tel. 261.1920
Rua Senador Dantas 19 s/205-6, RIO DE JANEIRO GB.
Tel. 232.07.32

CANADA
Renouf Publishing Company Limited,
2182 St. Catherine Street West,
MONTRÉAL, Que. H3H 1M7. Tel. (514)937.3519
OTTAWA, Ont. K1P 5A6, 61 Sparks Street

DENMARK – DANEMARK
Munksgaard Export and Subscription Service
35, Nørre Søgade
DK 1370 KØBENHAVN K. Tel. +45.1.12.85.70

FINLAND – FINLANDE
Akateeminen Kirjakauppa
Keskuskatu 1, 00100 HELSINKI 10. Tel. 65.11.22

FRANCE
Bureau des Publications de l'OCDE,
2 rue André-Pascal, 75775 PARIS CEDEX 16. Tel. (1) 524.81.67
Principal correspondant :
13602 AIX-EN-PROVENCE : Librairie de l'Université.
Tel. 26.18.08

GERMANY – ALLEMAGNE
OECD Publications and Information Center
4 Simrockstrasse 5300 BONN Tel. (0228) 21.60.45

GREECE – GRÈCE
Librairie Kauffmann, 28 rue du Stade,
ATHÈNES 132. Tel. 322.21.60

HONG-KONG
Government Information Services,
Publications/Sales Section, Baskerville House,
2/F., 22 Ice House Street

ICELAND – ISLANDE
Snaebjörn Jönsson and Co., h.f.,
Hafnarstraeti 4 and 9, P.O.B. 1131, REYKJAVIK.
Tel. 13133/14281/11936

INDIA – INDE
Oxford Book and Stationery Co. :
NEW DELHI-1, Scindia House. Tel. 45896
CALCUTTA 700016, 17 Park Street. Tel. 240832

INDONESIA – INDONÉSIE
PDIN-LIPI, P.O. Box 3065/JKT., JAKARTA, Tel. 583467

IRELAND – IRLANDE
TDC Publishers – Library Suppliers
12 North Frederick Street, DUBLIN 1 Tel. 744835-749677

ITALY – ITALIE
Libreria Commissionaria Sansoni :
Via Lamarmora 45, 50121 FIRENZE. Tel. 579751
Via Bartolini 29, 20155 MILANO. Tel. 365083
Sub-depositari :
Editrice e Libreria Herder,
Piazza Montecitorio 120, 00 186 ROMA. Tel. 6794628
Libreria Hoepli, Via Hoepli 5, 20121 MILANO. Tel. 865446
Libreria Lattes, Via Garibaldi 3, 10122 TORINO. Tel. 519274
La diffusione delle edizioni OCSE è inoltre assicurata dalle migliori librerie nelle città più importanti.

JAPAN – JAPON
OECD Publications and Information Center,
Landic Akasaka Bldg., 2-3-4 Akasaka,
Minato-ku, TOKYO 107 Tel. 586.2016

KOREA – CORÉE
Pan Korea Book Corporation,
P.O. Box n° 101 Kwangwhamun, SÉOUL. Tel. 72.7369

LEBANON – LIBAN
Documenta Scientifica/Redico,
Edison Building, Bliss Street, P.O. Box 5641, BEIRUT.
Tel. 354429 – 344425

MALAYSIA – MALAISIE
and/et **SINGAPORE - SINGAPOUR**
University of Malaysia Co-operative Bookshop Ltd.
P.O. Box 1127, Jalan Pantai Baru
KUALA LUMPUR. Tel. 51425, 54058, 54361

THE NETHERLANDS – PAYS-BAS
Staatsuitgeverij
Verzendboekhandel Chr. Plantijnstraat 1
Postbus 20014
2500 EA S-GRAVENAGE. Tel. nr. 070.789911
Voor bestellingen: Tel. 070.789208

NEW ZEALAND – NOUVELLE-ZÉLANDE
Publications Section,
Government Printing Office Bookshops:
AUCKLAND: Retail Bookshop: 25 Rutland Street,
Mail Orders: 85 Beach Road, Private Bag C.P.O.
HAMILTON: Retail Ward Street,
Mail Orders, P.O. Box 857
WELLINGTON: Retail: Mulgrave Street (Head Office),
Cubacade World Trade Centre
Mail Orders: Private Bag
CHRISTCHURCH: Retail: 159 Hereford Street,
Mail Orders: Private Bag
DUNEDIN: Retail: Princes Street
Mail Order: P.O. Box 1104

NORWAY – NORVÈGE
J.G. TANUM A/S Karl Johansgate 43
P.O. Box 1177 Sentrum OSLO 1. Tel. (02) 80.12.60

PAKISTAN
Mirza Book Agency, 65 Shahrah Quaid-E-Azam, LAHORE 3.
Tel. 66839

PHILIPPINES
National Book Store, Inc.
Library Services Division, P.O. Box 1934, MANILA.
Tel. Nos. 49.43.06 to 09, 40.53.45, 49.45.12

PORTUGAL
Livraria Portugal, Rua do Carmo 70-74,
1117 LISBOA CODEX. Tel. 360582/3

SPAIN – ESPAGNE
Mundi-Prensa Libros, S.A.
Castelló 37, Apartado 1223, MADRID-1. Tel. 275.46.55
Libreria Bastinos, Pelayo 52, BARCELONA 1. Tel. 222.06.00

SWEDEN – SUÈDE
AB CE Fritzes Kungl Hovbokhandel,
Box 16 356, S 103 27 STH, Regeringsgatan 12,
DS STOCKHOLM. Tel. 08/23.89.00

SWITZERLAND – SUISSE
OECD Publications and Information Center
4 Simrockstrasse 5300 BONN. Tel. (0228) 21.60.45
Local Agents/Agents locaux
Librairie Payot, 6 rue Grenus, 1211 GENÈVE 11. Tel. 022.31.89.50
Freihofer A.G., Weinbergstr. 109, CH-8006 ZÜRICH.
Tel. 01.3634282

THAILAND – THAILANDE
Suksit Siam Co., Ltd., 1715 Rama IV Rd,
Samyan, BANGKOK 5. Tel. 2511630

TURKEY – TURQUIE
Kültur Yayinlari Is-Türk Ltd. Sti.
Atatürk Bulvari No : 77/B
KIZILAY/ANKARA. Tel. 17 02 66
Dolmabahce Cad. No : 29
BESIKTAS/ISTANBUL. Tel. 60 71 88

UNITED KINGDOM – ROYAUME-UNI
H.M. Stationery Office, P.O.B. 569,
LONDON SE1 9NH. Tel. 01.928.6977, Ext. 410 or
49 High Holborn, LONDON WC1V 6 HB (personal callers)
Branches at: EDINBURGH, BIRMINGHAM, BRISTOL,
MANCHESTER, CARDIFF, BELFAST.

UNITED STATES OF AMERICA – ÉTATS-UNIS
OECD Publications and Information Center, Suite 1207,
1750 Pennsylvania Ave., N.W. WASHINGTON, D.C.20006 4582
Tel. (202) 724.1857

VENEZUELA
Libreria del Este, Avda. F. Miranda 52, Edificio Galipan,
CARACAS 106. Tel. 32.23.01/33.26.04/33.24.73

YUGOSLAVIA – YOUGOSLAVIE
Jugoslovenska Knjiga, Terazije 27, P.O.B. 36, BEOGRAD.
Tel. 621.992

Les commandes provenant de pays où l'OCDE n'a pas encore désigné de dépositaire peuvent être adressées à :
OCDE, Bureau des Publications, 2, rue André-Pascal, 75775 PARIS CEDEX 16.

Orders and inquiries from countries where sales agents have not yet been appointed may be sent to:
OECD, Publications Office, 2 rue André-Pascal, 75775 PARIS CEDEX 16.

64843-1-1982

OECD PUBLICATIONS, 2, rue André-Pascal, 75775 PARIS CEDEX 16 - No. 42207 1982
PRINTED IN FRANCE
(97 82 04 1) ISBN 92-64-12311-3

30 0005539 5